Random Permanents

Random Permanents

Yu. V. Borovskikh
Department of Applied Mathematics
Transport University
St. Petersburg, Russia

and

V.S. Korolyuk
Department of Probability Theory
Institute of Mathematics
Kiev, Ukraine

///VSP///
Utrecht, The Netherlands, 1994

VSP BV
P.O. Box 346
3700 AH Zeist
The Netherlands

First published in 1994

ISBN 90-6764-184-7

CIP-DATA KONINKLIJKE BIBLIOTHEEK, DEN HAAG

Borovskikh, Yuri V.

Random permanents / by Yuri V. Borovskikh and Vladimir S.
Korolyuk. - Utrecht : VSP
With index, ref.
ISBN 90-6764-184-7 bound
NUGI 811
Subject headings: matrix algebra

Printed in The Netherlands by Koninklijke Wöhrmann BV, Zutphen.

Contents

Preface

The theory of permanents of matrices is an interesting and promising approach to matrix algebra. In probability theory, the permanents of matrices with random elements are widely considered. In particular, when the elements of matrices can be regarded as counting random measures, the permanents of these matrices can be used while constructing an integral representation of U-statistics.

It is now evident that permanents are a highly peculiar and astonishing resource of mathematical research. This can be explained mainly by the fact that permanents are constructed by the use of a symmetrization procedure. It is thus quite natural that the application of permanents to the theory of symmetric statistics proves to be very promising.

Both in the theory of symmetric statistics and for U-statistics in particular, the variety of limit theorems have not yet been structurally formed.

At the same time, it is evident that the symmetrization property is principal. The determination of permanent random measures and the representation of symmetric statistics as functionals of symmetrization random measures with some deterministic kernels, make it possible to clarify the influence of the properties of a random measure upon the limiting results for symmetric statistics and also to study the influence of the characteristic structure of these kernels. This approach in the theory of symmetric statistics inspired the authors to investigate random permanents and their generating functions in detail.

New limiting results for random permanents are basically obtained by employing the algebraic and analytical properties of the permanents of sampling matrices and their generating functions. This notion allows the clarification of different schemes in the asymptotic analysis of symmetric statistics as the size of a sample n tends to infinity. Furthermore, the multiplicity of samples m may be fixed or increasing together with size n. The asymptotic analysis of symmetric statistics can be realized in the series scheme by use of normal approximation schemes and in the Poisson approximation as well.

It is worth noting that it is also possible to employ the infinitely-divisible approximation scheme. The authors intend to develop this scheme in future.

Within the framework of this approach, the theory of U statistics is a particular case of the theory of random permanents. This is a consequence of the above-mentioned fact that U-statistics are functionals of permanent random measures with some symmetric deterministic kernels.

The expansion of symmetric statistics over the orthogonal system of degenerate symmetric statistics is analogous to the expansion of stochastic multiple Ito–Wiener or Ito–Poisson functionals over Hermite or Charlier stochastic polynomials. Therefore the permanent theory of symmetric statistics can be considered as a statistical image of chaotic processes.

The research presented in this book was supported by the Fundamental Research Foundation of Sciences and Technology State Committee of Ukraine.

We appreciate the work of O.Adamenko and E.Cherkashin who kindly prepared the text ready for publication using personal computers of the Kiev Institute of Mathematics.

Introduction

The theory of permanents is a very interesting part of the matrix algebra. The permanent of a rectangular matrix A of $m \times n$ dimension ($m \leq n$):

$$A = [a_{ij};\ 1 \leq i \leq m,\ 1 \leq j \leq n]$$

is defined by the formula

$$\text{Per } A := \sum_{1 \leq j_1 \neq \cdots \neq j_m \leq n} \prod_{s=1}^{m} a_{s j_s}.$$

Every term in the sum is the product of elements of the matrix A, which are taken one by one from each row and each column.

A new area of investigations into probability theory is that of permanents of matrices with random elements. Particularly, when all rows of matrix A are equal and are given by simple sampling $\xi_1, \ldots, \xi_n$:

$$A_n^m(\xi) := [a_{ij} = \xi_j;\ 1 \leq i \leq m, 1 \leq j \leq n],$$

Per $A_n^m(\xi)$ is a symmetric polynomial of degree m. For permanents of matrices with random elements, the asymptotic problems of probability theory: the limit behaviour with increasing sample size n up to infinity, while m is fixed, or by increasing m together with n, is studied.

New perspectives arise with the introduction of a permanent random measure

$$\pi_n^\zeta(B_1, \ldots, B_m) := n^{-[m]} \text{Per } A_n^m(B_1, \ldots, B_m),$$

which is given by the permanent of the matrix

$$A_n^m(B_1, \ldots, B_m) := [a_{ij} = \zeta_j(B_i);\ 1 \leq i \leq m,\ 1 \leq j \leq n]$$

whose elements are independent random measures $\zeta_j(B_i)$ on measurable space $(X, \mathcal{X})$, $n^{[m]} = n(n-1) \cdots (n-m+1)$.

With the help of a permanent random measure, permanent symmetric statistics are determined

$$U_{nm}^\zeta(\Phi) := \int \cdots \int \Phi(x_1, \ldots, x_m) \pi_n^\zeta(dx_1, \ldots, dx_m)$$

with the kernel $\Phi : X^m \to R$.

In the particular case when $\zeta_j(B)$ are Dirac δ-measures, permanent symmetric statistics $U_{nm}^\zeta(\Phi)$ are U-statistics (Borovskikh et al. (1994)).

The introduction of a permanent random measure and the representation of permanent symmetric statistics as functionals of this symmetric random measure with a certain deterministic kernel allows clarification of the influence of the properties of permanent measures in limit results for symmetric statistics and, besides that, an explanation of the role of structural properties of the kernel of symmetric statistics.

First study the properties of permanents for a matrix whose dimension is $m \times n$ $(m \leq n)$. The main objects under investigation are permanents of matrices for which decomposition formulas are established, the generating function is derived, and shift decomposition formulas for the permanent of a sum of two matrices (chapter 1) are obtained.

Chapter 2 is devoted to the investigation of limiting behaviour, as $n \to \infty$, of the random permanents generated by the simple sample size of n (or the sample multiplicity of m) random variables in the series scheme under the Poisson approximation condition. The more complicated case, when m is increasing to infinity, is investigated in the following case $(m/n \to b, 0 < b < 1)$.

A permanent counting measure and permanent multiple functionals (chapter 3) are also examined.

Interesting results are obtained in chapter 4 where a normal approximation scheme is employed, in the case of fixed m. Here the relation between Hermite polynomials and the random permanents of sampling matrices are discovered. By virtue of theorems 4.4 and 4.5, it can be concluded that random permanents can be represented as a symmetric statistical image of Hermite polynomials.

The central point of this book occupies chapter 5, which is devoted to the investigation of the limit behaviour of permanent symmetric statistics represented as stochastic functionals from a deterministic kernel over the permanent random measure. The asymptotic analysis of symmetric statistics is realized in the normal and Poisson approximation schemes. It is worth noting that it is also possible to consider the infinitely-dimensional approximation scheme.

In chapter 6 some new possibilities for investigation of the mixture of Poisson and Gaussian approximation schemes are considered. Considering the weak convergence of permanent functionals, the investigation of the rate of convergence is natural and is presented in chapter 7.

In the final chapter 8 some perspectives are illustrated in the investigation of the permanents of sampling matrices with random elements as martingale differences, bootstrap samples and order statistics.

Chapter 1

Permanents of matrices

Permanents of rectangular matrices with real-valued (or complex-valued) elements are considered. The main objects under investigation are permanents of finite sampling matrices for which the decomposition formula is established and the generating function of permanents is derived.

The principal properties of permanents are the validity of the Laplace decomposition formula and the symmetric character of permanents of sampling matrices.

1.1 Definition and properties

The object under consideration is a matrix

$$A_n^m = [a_{ij}; 1 \leq i \leq m, 1 \leq j \leq n],$$

whose dimension is $m \times n$ $(m \leq n)$ and its elements a_{ij} are real-valued (or complex-valued) numbers.

The permanent of a matrix A_n^m is defined as a sum of products

$$\text{Per } A_n^m := \sum_{1 \leq j_1 \neq \cdots \neq j_m \leq n} \prod_{s=1}^{m} a_{sj_s}. \tag{1.1}$$

Each term in (1.1) is a product of the matrix elements taken one by one from each row and each column.

The number of terms in (1.1) is equal to the permanent of the matrix E_n^m with dimension $m \times n$ and all elements equal to 1 ($e_{ij} = 1$ for all i, j):

$$\begin{aligned} \text{Per } E_n^m &= n!/(n-m)! \\ &= n(n-1)\ldots(n-m+1) \\ &=: n^{[m]}. \end{aligned} \tag{1.2}$$

Example 1.1. The *sampling matrix with identity rows* is determined by the set of numbers $a_1, a_2, \ldots, a_n$:

$$A_n^m(a) := [a_{ij} = a_j; 1 \le i \le m, 1 \le j \le n]. \tag{1.3}$$

The permanent of a sampling matrix is a symmetric polynomial $P_n^m(a)$ of n variables $a := (a_1, a_2, \ldots, a_n)$ of degree m:

$$P_n^m(a) := \sum_{1 \le j_1 < \cdots < j_m \le n} \prod_{s=1}^{m} a_{j_s}, \tag{1.4}$$

with factor $m!$:

$$\text{Per } A_n^m(a) = m! P_n^m(a). \tag{1.5}$$

This fact allows the investigation of symmetric polynomials as permanents of a sampling matrix.

1.2 Laplace formula

The main purpose is to investigate permanents of sampling matrices.

Introduce sets of sequences of increasing positive numbers

$$Q_r = \{q = (q_1, \ldots, q_r) : \ 1 \le q_1 < \ldots < q_r \le n\}.$$

Let $A_n^m\,[p|q]$, $p \in Q_r$, $q \in Q_s$ be the submatrices of A_n^m situated on the intersection of the $(p_1, \ldots, p_r)$th rows and $(q_1, \ldots, q_s)$th columns; and let $A_n^m\,(p|q)$, $p \in Q_r$, $q \in Q_s$ be the submatrices which are obtained from A_n^m by deleting rows with numbers $p_1, \ldots, p_r$ and columns with numbers $q_1, \ldots, q_s$.

The submatrices $A_n^m\,[p|q]$ and $A_n^m\,(p|q)$ have dimensions $r \times s$ and $(m-r) \times (n-s)$ respectively:

$$A_n^m\,[p|q] := [a_{p_i q_j}; 1 \le i \le r, 1 \le j \le s],$$

$$A_n^m\,(p|q) := [a_{p'_i q'_j}; 1 \le i \le m-r, 1 \le j \le n-s, p'_i \notin p, q'_j \notin q].$$

The *Laplace formula* for a permanent has the following form

$$\text{Per } A_n^m = \sum_{q \in Q_r} \text{Per } A_n^m[p|q] \text{ Per } A_n^m(p|q) \tag{1.6}$$

for all $p \in Q_r$, $1 \le r \le m \le n$. In the specific case when $r = m$

$$\text{Per } A_n^m = \sum_{q \in Q_m} \text{Per} A_n^m[-|q], \tag{1.7}$$

where $A_n^m[-|q]$ is the square $m \times m$ matrix with elements from the $(q_1, \ldots, q_m)$-th columns, i.e.,

$$A_n^m[-|q] := [a_{iq_j}; \ 1 \le i \le m, 1 \le j \le m].$$

The specific case of $r = 1$ leads to the Laplace formula

$$\text{Per } A_n^m = \sum_{j=1}^{n} a_{ij} \text{Per } A_n^m(i|j).$$

Formulas (1.4)–(1.5) for the permanent of a sampling matrix can be established by application of the Laplace formula (1.7) in which the square $m \times m$-matrix $A_n^m[-|q]$, with equal rows, has the permanent

$$\text{Per } A_n^m[-|q] = m! \prod_{s=1}^{m} a_{q_s}.$$

Combine this equality with (1.7) to obtain the formulas (1.4)–(1.5) in example 1.1.

The application of the Laplace formula is illustrated by the following examples.

Example 1.2. The square $n \times n$-matrix

$$\bar{A}_n^n := \begin{bmatrix} A_n^m \\ E_n^{n-m} \end{bmatrix},$$

where E_n^{n-m} stands for the $(n-m) \times n$-matrix with all elements equal to 1, has the permanent

$$\text{Per } \bar{A}_n^n = (n-m)! \text{ Per } A_n^m. \tag{1.8}$$

Indeed, applying the Laplace formula (1.6) with $r = m$ and $p = (1, 2, \ldots, m)$,

$$\begin{aligned} \text{Per } \bar{A}_n^n &= \sum_{q \in Q_m} \text{Per } A_n^m[-|q] \text{ Per } E_{n-m}^{n-m} \\ &= (n-m)! \sum_{q \in Q_m} \text{Per } A_n^m[-|q] \quad \text{using (1.2)} \\ &= (n-m)! \text{ Per } A_n^m \quad \text{by the Laplace formula (1.7).} \end{aligned}$$

Example 1.3. The augmented matrix with dimension $m' \times n$ $(m \leq m' \leq n)$

$$\bar{A}_n^{m'} := \begin{bmatrix} A_n^m \\ E_n^{m'-m} \end{bmatrix},$$

has the permanent

$$\text{Per } \bar{A}_n^{m'} = \frac{(n-m)!}{(n-m')!} \text{ Per } A_n^m. \tag{1.9}$$

Indeed, as mentioned above, by virtue of the Laplace formula (1.6) with $r = m$ and $p = (1, 2, \ldots, m)$,

$$\begin{aligned} \text{Per } \bar{A}_n^{m'} &= \sum_{q \in Q_m} \text{Per } A_n^m[-|q] \text{ Per } E_{n-m}^{m'-m} \\ &= (n-m)^{[m'-m]} \sum_{q \in Q_m} \text{Per } A_n^m[-|q] \quad \text{using (1.2)} \end{aligned}$$

$$= (n-m)^{[m'-m]}\text{Per } A_n^m$$

$$= \frac{(n-m)!}{(n-m')!}\text{ Per } A_n^m \quad \text{by (1.7).}$$

1.3 Decomposition formula

The principal advantage of employing the Laplace formula is the possibility of obtaining a decomposition for the permanents of multiple sampling matrices. The *multiple sampling matrix* $A_n^m(a_1, \ldots, a_k)$ is defined as

$$A_n^m(a_1,\ldots,a_k) := \begin{bmatrix} A_n^{m_1}(a_1) \\ \cdots \\ A_n^{m_k}(a_k) \end{bmatrix}, \tag{1.10}$$

where $m_1 + \cdots + m_k = m$ and $A_n^{m_s}(a_s)$ are sampling matrices with dimension $m_s \times n$ and elements $a_s = (a_{s1}, \ldots, a_{sn})$.

First consider the special case of $k = 2$, $m_1 + m_2 = m$. Applying the Laplace formula (1.6) to the two-sampling matrix $A_n^m(a_1, a_2)$ by setting $p = (1, 2, \ldots, m_1)$, obtain

$$\text{Per } A_n^m(a_1,a_2) = \sum_{q\in Q_{m_1}} \text{Per } A_n^{m_1}[-|q] \text{ Per } A_n^{m_2}(-|q),$$

where $A_n^{m_1}[-|q]$ is a sampling square matrix whose dimension is $m_1 \times m_1$ with $q_1, \ldots, q_{m_1}$ columns. Hence

$$\text{Per } A_n^{m_1}[-|q] = m_1! \prod_{s=1}^{m_1} a_{1q_s}.$$

Further, $A_n^{m_2}(-|q)$ is a sampling matrix whose dimension is $m_2 \times (n - m_1)$ and the columns $q_1, \ldots, q_{m_1}$ are deleted.

Hence the permanent of this matrix can be represented as

$$\text{Per } A_n^{m_2}(-|q) = m_2! \sum_{q'\in Q_{m_2}^{n-m_1}(q)} \prod_{s=1}^{m_2} a_{2q'_s},$$

where the set $Q_{m_2}^{n-m_1}(q)$ contains the increasing sequence of positive integer numbers $q' = (q'_1, \ldots, q'_{m_2})$, with $q_1, \ldots, q_{m_1}$ being deleted:

$$Q_{m_2}^{n-m_1}(q) := \{q' = (q'_1,\ldots,q'_{m_2}) :$$

$$1 \le q'_1 < \ldots < q'_{m_2} \le n;\ q'_s \ne q_l,\ 1 \le l \le m_1\}.$$

Combine these representations to obtain the decomposition formula for the permanent of a two-sampling matrix

$$\text{Per } A_n^m(a_1,a_2) = m_1!m_2! \sum_{q\in Q_{m_1}} \prod_{s=1}^{m_1} a_{1q_s} \sum_{q'\in Q_{m_2}^{n-m_1}(q)} \prod_{s=1}^{m_2} a_{2q'_s}. \tag{1.11}$$

In the general case of $k \geq 2$ introduce the sets of sequences of increasing positive integers

$$Q_{m_s}^{n-\bar{m}_s}(\tilde{q}^{(s-1)}) :=$$

$$:= \{q = (q_1,\ldots,q_{m_s}) : \ 1 \leq q_1 < \ldots < q_{m_s} \leq n, q_s \neq q_l^{(r)},$$

$$1 \leq l \leq m_r, \ 1 \leq r \leq s-1\},$$

where

$$\bar{m}_s := m - \sum_{i=s}^{k} m_i, \quad \bar{m}_1 = 0,$$

$$\bar{m}_k = m_1 + \cdots + m_{k-1},$$

$$\tilde{q}^{(s)} := (q^{(1)},\ldots,q^{(s)}).$$

The sets $Q_{m_s}^{n-\bar{m}_s}(\tilde{q}^{(s-1)})$ contain collections of increasing positive integers $(1,2,\ldots,n)$ with the deleted collections $q^{(r)}, \ 1 \leq r \leq s-1, \ s = 2,\ldots,k$.

Theorem 1.1. *The permanent of multiple sampling matrix* (1.10) *has decomposition of the following form*

$$\text{Per } A_n^m(a_1,\ldots,a_k) = \prod_{s=1}^{k} m_s! \sum_{q^{(1)}\in Q_{m_1}} \prod_{s=1}^{m_1} a_{1q_s^{(1)}} \sum_{q^{(2)}\in Q_{m_2}^{n-\bar{m}_2}(q^{(1)})} \prod_{s=1}^{m_2} a_{2q_s^{(2)}} \times$$

$$\cdots \times \sum_{q^{(k)}\in Q_{m_k}^{n-\bar{m}_k}(\tilde{q}^{(k-1)})} \prod_{s=1}^{m_k} a_{kq_s^{(k)}}. \tag{1.12}$$

Proof. By application of Laplace formula (1.6) to the permanent of matrix (1.10)

$$\text{Per } A_n^m(a_1,\ldots,a_k) = m_1! \sum_{q^{(1)}\in Q_{m_1}} \prod_{s=1}^{m_1} a_{1q_s^{(1)}} \text{Per } A_n^{m-\bar{m}_2}(p^{(1)}|q^{(1)}),$$

where the matrix $A_n^{m-\bar{m}_2}(p^{(1)}|q^{(1)})$ is obtained from $A_n^m(a_1,\ldots,a_k)$ by deleting the rows with numbers $p^{(1)} = (1,2,\ldots,m_1)$ and columns with numbers $q^{(1)} = (q_1^{(1)},\ldots q_{m_1}^{(1)})$.

Thus the size of the matrix $A_n^{m-\bar{m}_2}(p^{(1)}|q^{(1)})$ is $(m - \bar{m}_2) \times (n - m_1)$. On the second step, by application of the Laplace formula (1.6) the permanent of the matrix $A_n^{m-\bar{m}_2}(p^{(1)}|q^{(1)})$

$$\text{Per } A_n^{m-\bar{m}_2}(p^{(1)}|q^{(1)}) = m_2! \sum_{q^{(2)} \in Q_{m_2}^{n-\bar{m}_2}(q^{(1)})} \prod_{s=1}^{m_2} a_{2q_s^{(2)}} \text{Per } A_n^{m-\bar{m}_3}(p^{(2)}|q^{(2)}),$$

where the matrix $A_n^{m-\bar{m}_3}(p^{(2)}|q^{(2)})$ with dimension $(m - \bar{m}_3) \times (n - m_1 - m_2)$ is obtained from $A_n^{m-\bar{m}_2}(p^{(1)}|q^{(1)})$ by deleting the rows with numbers

$$p^{(2)} = (m_1 + 1, \ldots, m_1 + m_2)$$

and the columns with numbers

$$q^{(2)} = (q_1^{(2)}, \ldots, q_{m_2}^{(2)}).$$

Hence the matrix $A_n^{m-\bar{m}_3}(p^{(2)}|q^{(2)})$ is obtained from $A_n^m(a_1, \ldots, a_k)$ by deleting the rows with numbers $1, 2, \ldots, m_1 + m_2$ and the columns with numbers $\tilde{q}^{(2)} = (q_1^{(1)}, \ldots, q_{m_1}^{(1)}, q_1^{(2)}, \ldots, q_{m_2}^{(2)})$.

If this procedure is repeated $k - 1$ times the decomposition formula (1.12) is obtained.

Corollary 1.1. *The permanent of a multiple sampling matrix with $k = m$, $m_1 = \ldots = m_k = 1$ can be decomposed in the following way*

$$\text{Per } A_n^m(a_1, \ldots, a_m) = \sum_{q_1=1}^{n} a_{1q_1} \sum_{q_2 \in Q_1^{n-1}(q_1)} a_{2q_2} \cdots \sum_{q_m \in Q_1^{n-m+1}(q_1, \ldots, q_{m-1})} a_{mq_m}.$$

1.4 Generating function

First study the sampling matrices presented in example 1.1.

Lemma 1.1. *The generating function of the permanents of sampling matrices $A_n^m(a), 0 \leq m \leq n$, is given by the formula*

$$\begin{aligned} \Pi_n(z; a) &:= \sum_{m=0}^{n} \frac{z^m}{m!} \text{Per } A_n^m(a) \\ &= \prod_{j=1}^{n} (1 + z a_j). \end{aligned} \tag{1.13}$$

Proof. This formula is a simple consequence of the well-known identity

$$\prod_{j=1}^{n}(1+za_j)=\sum_{m=0}^{n}z^m P_n^m(a), \tag{1.14}$$

where $P_n^m(a)$ are symmetric m-degree polynomials of n variables and $a=(a_1,\ldots,a_n)$ (see (1.4) and equality (1.5)).

For $n=1$, the identity (1.14) is trivial. In order to prove (1.14) for all $n\geq 1$, use the following identities for symmetric polynomials with $m\leq n$

$$P_{n+1}^m(a)=P_n^m(a)+a_{n+1}P_n^{m-1}(a),$$

$$P_{n+1}^{n+1}(a)=a_{n+1}P_n^n(a).$$

Now evaluate

$$\Pi_{n+1}(z;a):=\sum_{m=0}^{n+1}z^m P_{n+1}^m(a)=$$

$$=1+\sum_{m=1}^{n}z^m P_{n+1}^m(a)+z^{n+1}P_{n+1}^{n+1}(a)=$$

$$=1+\sum_{m=1}^{n}z^m P_n^m(a)+a_{n+1}\sum_{m=1}^{n}z^m P_n^{m-1}(a)+z^{n+1}a_{n+1}P_n^n(a)=$$

$$=\Pi_n(z;a)+za_{n+1}\sum_{m=0}^{n-1}z^m P_n^m(a)+za_{n+1}z^n P_n^n(a)=$$

$$=\Pi_n(z;a)+za_{n+1}\sum_{m=0}^{n}z^m P_n^m(a).$$

As a result the identity

$$\Pi_{n+1}(z;a)=\Pi_n(z;a)(1+za_{n+1})$$

is obtained.

As an induction hypothesis, assume that (1.13) holds for a fixed number $n\geq 1$, then the last identity yields the result of lemma 1.1 for all $n\geq 1$.

Theorem 1.2. *The generating function of permanents of multiple sampling matrices $A_n^m(a_1,\ldots,a_k)$ is given by the formula*

$$\begin{aligned}\Pi_n(z;a_1,\ldots,a_k) &:= \sum_{m=0}^{n}\sum_{m_1+\cdots+m_k=m}\prod_{s=1}^{k}\frac{z_s^{m_s}}{m_s!}\mathrm{Per}A_n^m(a_1,\ldots,a_k)\\ &= \prod_{j=1}^{n}[1+\sum_{s=1}^{k}z_s a_{sj}].\end{aligned} \tag{1.15}$$

Proof. First examine the case of $k=2, m=m_1+m_2$, by use of the decomposition formula (1.11).

$$\begin{aligned}\Pi_n(z;a_1,a_2) &= \sum_{m=0}^{n}\sum_{m_1+m_2=m}\frac{z_1^{m_1}z_2^{m_2}}{m_1!m_2!}\mathrm{Per}A_n^m(a_1,a_2)\\ &= \sum_{m_1=0}^{n} z_1^{m_1}\sum_{q\in Q_{m_1}}\prod_{s=1}^{m_1}a_{1q_s}\sum_{m_2=0}^{n-m_1}z_2^{m_2}\sum_{q'\in Q_{m_2}^{n-m_1}(q)}\prod_{s=1}^{m_2}a_{2q'_s}.\end{aligned}$$

The last sums over q' are symmetric m_2-degree polynomials $P_{n-m_1}^{m_2}(\tilde{a}_2;q)$ of the variables $\tilde{a}_2=(a_{2j},1\le j\le n, j\notin q)$.

By the identity (1.14)

$$\sum_{m_2=0}^{n-m_1} z_2^{m_2}P_{n-m_1}^{m_2}(\tilde{a}_2;q)=\prod_{j=1}^{n}(1+z_2a_{2j})\prod_{s=1}^{m_1}(1+z_2a_{2q'_s})^{-1}.$$

Combine the last two equalities to get

$$\Pi_n(z;a_1,a_2)=\sum_{m_1=0}^{n} z_1^{m_1}\sum_{q\in Q_{m_1}}\prod_{s=1}^{m_1}a_{1q_s}(1+z_2a_{2q'_s})^{-1}\Pi_n(z_2;a_2).$$

These sums, over collections q, are the symmetric m_1-degree polynomials $P_n^{m_1}(\tilde{a}_1)$ of the variables $\tilde{a}_1:=(a_{1j}(1+z_2a_{2j})^{-1};1\le j\le n)$.

By virtue of the identities (1.14) and (1.13)

$$\begin{aligned}\Pi_n(z;a_1,a_2)&=\prod_{j=1}^{n}[1+z_1a_{1j}(1+z_2a_{2j})^{-1}]\Pi_n(z_2;a_2)\\ &=\prod_{j=1}^{n}[1+z_1a_{1j}+z_2a_{2j}].\end{aligned}$$

To prove theorem 1.2 in the general case when $k\ge 2$, employ the decomposition formula (1.12).

$$\begin{aligned}&\Pi_n(z_1,\ldots,z_k)=\\ &=\sum_{0\le m_1+\cdots+m_k\le n}\ \sum_{q^{(1)}\in Q_{m_1}}\prod_{s=1}^{m_1}z_1a_{1q_s^{(1)}}\times\cdots\times\\ &\times\sum_{q^{(k)}\in Q_{m_k}^{n-m_k}(\bar{q}^{(k-1)})}\prod_{s=1}^{m_k}z_ka_{kq_s^{(k)}}.\end{aligned}\qquad(1.16)$$

The summation is undertaken with respect to the set of indices $0\le m_1+\cdots+m_k\le n$, $0\le m_j\le n$, $j=1,\ldots,k$, and it is given more detail. Then obtain

$$\Pi_n(z_1,\ldots,z_k)=\sum_{m_1=0}^{n}z_1^{m_1}\sum_{q^{(1)}\in Q_{m_1}}\prod_{s=1}^{m_1}a_{1q_s^{(1)}}\times$$

$$\times \sum_{m_2=0}^{n-m_1} z_2^{m_2} \sum_{q^{(2)}\in Q_{m_2}^{n-\bar{m}_2}(\bar{q}^{(1)})} \prod_{s=1}^{m_2} a_{2q_s^{(2)}} \times \ldots \times$$

$$\times \sum_{m_{k-1}=0}^{n-\bar{m}_{k-1}} z_{k-1}^{m_{k-1}} \sum_{q^{(k-1)}\in Q_{m_{k-1}}^{n-\bar{m}_{k-1}}(\bar{q}^{(k-2)})} \prod_{s=1}^{m_{k-1}} a_{k-1,q_s^{(k-1)}} \times$$

$$\times \sum_{m_k=0}^{n-\bar{m}_k} z_k^{m_k} \sum_{q^{(k)}\in Q_{m_k}^{n-\bar{m}_k}(\bar{q}^{(k-1)})} \prod_{s=1}^{m_k} a_{kq_s^{(k)}}. \tag{1.17}$$

For the last sum, with respect to m_k in the right-hand side of (1.17) in accordance with (1.14), write

$$\sum_{m_k=0}^{n-\bar{m}_k} z^{m_k} \sum_{q^{(k)}\in Q_{m_k}^{n-\bar{m}_k}(\bar{q}^{(k-1)})} \prod_{s=1}^{m_k} a_{kq_s^{(k)}} =$$

$$= \prod_{j=1}^{n}(1 + z_k a_{kj}) \prod_{r=1}^{k-1}\prod_{s=1}^{m_r}(1 + z_k a_{kq_s^{(r)}})^{-1}. \tag{1.18}$$

With substitution of (1.18) into (1.17) it is found that

$$\Pi_n(z_1, \ldots, z_k) = \prod_{j=1}^{n}(1 + z_k a_{kj}) \sum_{m_1=0}^{n} z_1^{m_1} \sum_{q^{(1)}\in Q_{m_1}}$$

$$\times \prod_{s=1}^{m_1} a_{1q_s^{(1)}}(1 + z_k a_{kq_s^{(1)}})^{-1} \times \cdots \times$$

$$\times \sum_{m_{k-1}=0}^{n-\bar{m}_{k-1}} z_{k-1}^{m_{k-1}} \sum_{q^{(k-1)}\in Q_{m_{k-1}}^{n-\bar{m}_{k-1}}(\bar{q}^{(k-2)})}$$

$$\times \prod_{s=1}^{m_{k-1}} a_{k-1,q_s^{(k-1)}}(1 + z_k a_{kq_s^{(k-1)}})^{-1}. \tag{1.19}$$

Apply (1.14) to the last sum, with respect to m_{k-1} in the right-hand side of (1.19), again:

$$\sum_{m_{k-1}=0}^{n-\bar{m}_{k-1}} z_{k-1}^{m_{k-1}} \sum_{q^{(k-1)}\in Q_{m_{k-1}}^{n-\bar{m}_{k-1}}(\bar{q}^{(k-2)})} \prod_{s=1}^{m_{k-1}} a_{k-1,q_s^{(k-1)}}(1 + z_k a_{kq_s^{(k-1)}})^{-1} =$$

$$= \prod_{j=1}^{n}(1 + z_{k-1}a_{k-1,j}(1 + z_k a_{kj})^{-1})\times$$

$$\times \prod_{r=1}^{k-2}\prod_{s=1}^{m_r} \left(1 + z_{k-2}a_{k-2,q_s^{(r)}}\,(1 + z_k a_{kq_s^{(r)}})^{-1}\right)^{-1} =$$

$$= \prod_{j=1}^{n}(1 + z_k a_{kj} + z_{k-1}a_{k-1,j}) \prod_{r=1}^{k-2}\prod_{s=1}^{m_r}(1 + z_k a_{kq_s^{(r)}} +$$

$$+ z_{k-1}a_{k-1,q_s^{(r)}})^{-1} \prod_{j=1}^{n}(1 + z_k a_{kj})^{-1} \prod_{r=1}^{k-2}\prod_{s=1}^{m_r}(1 + z_k a_{kq_s^{(r)}}). \tag{1.20}$$

Further substitute (1.20) into (1.19) and reduce some co-factors. Then apply (1.14) $(k-2)$ times. The result of this method of calculation gives (1.15).

1.5 Shift decomposition of permanents

Consider a sampling matrix $A_n^m(1+a)$ with elements $a_{ij} = 1 + a_j$, $1 \le i \le m$, $1 \le j \le n$.

Theorem 1.3. *The following shift decomposition is true*

$$\mathrm{Per}A_n^m(1+a) = \frac{m!}{(n-m)!}\sum_{c=0}^{m}\frac{(n-c)!}{(m-c)!}\mathrm{Per}A_n^c(a)/c!. \tag{1.21}$$

Using the notation (1.2) *this formula can be transformed as*

$$\mathrm{Per}A_n^m(1+a) = n^{[m]}\sum_{c=0}^{m}\frac{m^{[c]}}{n^{[c]}}\mathrm{Per}A_n^c(a)/c!. \tag{1.22}$$

Proof. By formulas (1.4) and (1.5),

$$\begin{aligned}\mathrm{Per}A_n^m(1+a) &= m!P_n^m(1+a)\\ &= m!\sum_{q\in Q_m}\prod_{s=1}^{m}(1+a_{q_s}).\end{aligned} \tag{1.23}$$

Employing formulas (1.13) and (1.14)

$$\prod_{s=1}^{m}(1+a_{q_s}) = \sum_{c=0}^{m}\mathrm{Per}A_n^c[-|q]/c!, \tag{1.24}$$

where $A_n^c[-|q]$ are matrices with dimension $c \times m$ obtained from $A_n^c(a)$ with $q_1, \ldots, q_c$ columns.

By virtue of formula (1.8), for the permanent of the augmented matrix

$$\mathrm{Per}A_n^c[-|q] = \mathrm{Per}\bar{A}_n^m[-|q]/(m-c)!. \tag{1.25}$$

From formula (1.9) in example 1.3

$$\mathrm{Per}\bar{A}_n^m(a) = \frac{(n-c)!}{(n-m)!}\mathrm{Per}A_n^c(a). \tag{1.26}$$

Now putting (1.24) and (1.25) into (1.23) obtain

$$\mathrm{Per}A_n^m(1+a) = m!\sum_{c=0}^{m}\frac{1}{c!(m-c)!}\sum_{q\in Q_m}\mathrm{Per}\bar{A}_n^m[-|q]. \tag{1.27}$$

The Laplace formula (1.7) yields

$$\sum_{q\in Q_m}\mathrm{Per}\bar{A}_n^m[-|q] = \mathrm{Per}\bar{A}_n^m(a). \tag{1.28}$$

The relations (1.26)-(1.28) together imply (1.21).

Using the generating function of permanents (1.13) it is possible to obtain another proof of theorem 1.3. By the Cauchy formula and (1.13)

$$\mathrm{Per}\ A_n^m(1+a) = \frac{m!}{2\pi i}\oint\frac{dz}{z^{m+1}}\prod_{j=1}^{n}(1+z(1+a_j)), \tag{1.29}$$

where the integration is taken over the circle which contains the point $z=0$. Represent the integrand function in the form

$$\prod_{j=1}^{n}(1+z(1+a_j)) = (1+z)^n\prod_{j=1}^{n}\left(1+a_j\frac{z}{1+z}\right).$$

According to formula (1.13)

$$\prod_{j=1}^{n}\left(1+a_j\frac{z}{1+z}\right) = \sum_{c=0}^{n}\left(\frac{z}{1+z}\right)^c \mathrm{Per}\ A_n^c(a)/c!.$$

Producing

$$\prod_{j=1}^{n}(1+z(1+a_j)) = \sum_{c=0}^{n} z^c(1+z)^{n-c}\ \mathrm{Per}\ A_n^c(a)/c!. \tag{1.30}$$

Substitute representation (1.30) into (1.29) and obtain

$$\begin{aligned}\mathrm{Per}\ A_n^m(1+a) \ &= \ \sum_{c=0}^{n}\frac{m!}{c!}\mathrm{Per}\ A_n^c(a)\ \times \\ &\times\left\{\frac{1}{2\pi i}\oint\frac{dz}{z^{m-c+1}}(1+z)^{n-c}\right\}.\end{aligned}$$

Here

$$\frac{1}{2\pi i}\oint\frac{dz}{z^{m-c+1}}(1+z)^{n-c} = \begin{cases}\dfrac{(n-c)!}{(m-c)!(n-m)!}, & \text{if}\quad 0\le c\le m, \\[2ex] 0, & \text{if}\quad c>m.\end{cases} \tag{1.31}$$

So the formula (1.29) is transformed to (1.21).

Corollary 1.2. *Let*

$$A_n^m(h+a) = [a_{ij} = h + a_j;\ 1 \le i \le m, 1 \le j \le n],\ h \neq 0.$$

Then

$$\text{Per } A_n^m(h+a) = n^{[m]} \sum_{c=0}^{m} h^{m-c} n^{-[c]} m^{[c]} \text{Per} A_n^c(a)/c!. \tag{1.32}$$

Proof. Using the property of the permanent

$$\text{Per } A_n^m(h+a) = h^m \text{Per} A_n^m(1 + a/h),$$

and with substitution of a/h instead of a in (1.22), obtain (1.32).

The shift decomposition can be extended to a multiple sampling matrix, which is defined as in (1.10)

$$A_n^m(1+a_1, \dots, 1+a_k) = \begin{bmatrix} A_n^{m_1}(1+a_1) \\ \cdots \\ A_n^{m_k}(1+a_k) \end{bmatrix},$$

where $m_1 + \cdots + m_k = m$ and $A_n^{m_s}(1+a_s)$ are sampling matrices with dimension $m_s \times n$ and elements $1 + a_s = (1 + a_{sj}, 1 \le j \le n),\ 1 \le s \le k$.

Theorem 1.4. *The following shift decomposition is true*

$$n^{-[m]} \text{Per } A_n^m(1+a_1, \dots, 1+a_k) =$$

$$= \sum_{c=0}^{m} n^{-[c]} \sum_{r_1+\cdots+r_k=c} \prod_{s=1}^{k} m_s^{[r_s]} \text{Per } A_n^c(a_1, \dots, a_k) / \prod_{s=1}^{k} r_s!. \tag{1.33}$$

Proof. This theorem can be proved by the scheme of the second proof of theorem 1.3 and using the Cauchy formula

$$\text{Per } A_n^m(1+a_1, \dots, 1+a_k) =$$

$$= m_1! \dots m_k! \frac{1}{(2\pi i)^k} \oint \cdots \oint \prod_{s=1}^{k} \frac{dz_s}{z_s^{m_s+1}} \prod_{j=1}^{n} \left(1 + \sum_{s=1}^{k} z_s(1+a_{sj})\right),$$

where each integral is taken over the circle with small radius which contains the point $z_s = 0$. Instead of (1.31) it is necessary to use the following formula

$$\frac{1}{(2\pi i)^k} \oint \cdots \oint \prod_{s=1}^{k} \frac{dz_s}{z_s^{m_s - r_s + 1}} (1 + \sum_{s=1}^{k} z_s)^{n-c} =$$

$$= (n-c)!/(n-m)! \prod_{s=1}^{k} (m_s - r_s)!$$

for $0 \le c \le m = m_1 + \cdots + m_k$.

Corollary 1.3. *For the permanent of the multiple sampling matrix*

$$A_n^m(h_1 + a_1, \ldots, h_k + a_k) = \begin{bmatrix} A_n^{m_1}(h_1 + a_1) \\ \cdots \\ A_n^{m_k}(h_k + a_k) \end{bmatrix}$$

the following shift decomposition formula is valid:

$$n^{-[m]} \prod_{s=1}^{k} h_s^{-m_s} \text{Per } A_n^m(h_1 + a_1, \ldots, h_k + a_k) =$$

$$= \sum_{c=0}^{m} n^{-[c]} \sum_{r_1+\ldots+r_k=c} \prod_{s=1}^{k} h_s^{-r_s} m_s^{[r_s]} \text{ Per } A_n^c(a_1, \ldots, a_k) / \prod_{s=1}^{k} r_s!. \tag{1.34}$$

Proof is analogous to that of corollary 1.2.

Even more interesting is the shift decomposition of a permanent of the sum of two sampling matrices.

Theorem 1.5. *For the sum of two sampling matrices*

$$A_n^m(b + a) = [a_{ij} = b_j + a_j;\ 1 \le i \le m,\ 1 \le j \le n]$$

the shift decomposition of the permanent is valid

$$\text{Per } A_n^m(b + a) = \sum_{c=0}^{m} m^{[c]} \sum_{1 \le q_1 \ne \cdots \ne q_c \le n} \prod_{s=1}^{c} a_{q_s} \text{Per } A_{n-c}^{m-c}(\bar{b}; q)/c!, \tag{1.35}$$

where

$$A_{n-c}^{m-c}(\bar{b}; q) = [a_{ij} = b_j;\ 1 \le i \le m - c, 1 \le j \le n, j \ne q_1, \ldots, q_c].$$

Proof. According to the Cauchy formula and (1.13),

$$\text{Per } A_n^m(b + a) = \frac{m!}{2\pi i} \oint \frac{dz}{z^{m+1}} \prod_{j=1}^{n} (1 + z(b_j + a_j)). \tag{1.36}$$

The product in (1.36) can be represented in the following form

$$\prod_{j=1}^{n} (1 + z(b_j + a_j)) = \prod_{j=1}^{n} (1 + zb_j) \prod_{j=1}^{n} (1 + za_j/(1 + zb_j)).$$

According to formula (1.13)

$$\prod_{j=1}^{n} (1 + za_j/(1 + zb_j)) = \sum_{c=0}^{n} \frac{z^c}{c!} \sum_{1 \le j_1 \ne \ldots \ne j_c \le n} \prod_{s=1}^{c} a_{j_s}(1 + zb_{j_s})^{-1}.$$

Consequently, the permanent (1.36) has the following representation

$$\text{Per } A_n^m(b+a) = \sum_{c=0}^{n} \frac{m!}{c!} \sum_{1\le j_1\ne\cdots\ne j_c\le n} \prod_{s=1}^{c} a_{j_s} \times$$

$$\times \left\{ \frac{1}{2\pi i} \oint \frac{dz}{z^{m-c+1}} \prod_{j=1}^{n}(1+zb_j) \prod_{s=1}^{c}(1+zb_{j_s})^{-1} \right\}, \tag{1.37}$$

where

$$\prod_{j=1}^{n}(1+zb_j)\prod_{s=1}^{c}(1+zb_{j_s})^{-1} = \prod_{j=1}^{n-c}(1+z\bar{b}_j).$$

Notice that

$$\{\bar{b}_1, \dots, \bar{b}_{n-c}\} = \{b_1, \dots, b_n\} \backslash \{b_{j_1}, \dots, b_{j_c}\}.$$

But it is obvious that

$$\frac{1}{2\pi i} \oint \frac{dz}{z^{m-c+1}} \prod_{j=1}^{n-c}(1+z\bar{b}_j) = \begin{cases} \text{Per} A_{n-c}^{m-c}(\bar{b})/(m-c)!\,, & \text{if } \; 0 \le c \le m \\ 0\,, & \text{if } \quad c > m. \end{cases}$$

Put the last two formulas in (1.37) to get (1.35).

Corollary 1.4. *Let the sequence $b_1, b_2, \dots, b_n$ consist of zeros and units and*

$$\nu_n(b) = b_1 + b_2 + \ldots + b_n \tag{1.38}$$

is the number of units in this sequence. Then there is the shift decomposition

$$\text{Per} A_n^m(b+a) = \sum_{c=0}^{m} m^{[c]} \sum_{1\le j_1\ne\ldots\ne j_c\le n} \prod_{s=1}^{c} a_{j_s} \nu_{n-c}^{[m-c]}(\bar{b})/c!, \tag{1.39}$$

where, according to the definition of $\bar{b}$,

$$\nu_{n-c}(\bar{b}) = \nu_n(b) - b_{j_1} - \ldots - b_{j_c}.$$

Consider the sum of two multiple sampling matrices

$$A_n^m(b_1+a_1, \dots, b_k+a_k) = \begin{bmatrix} A_n^{m_1}(b_1+a_1) \\ \cdots \\ A_n^{m_k}(b_k+a_k) \end{bmatrix},$$

where

$$A_n^{m_s}(b_s+a_s) = [a_{ij} = b_{sj} + a_{sj}; 1 \le i \le m_s, 1 \le j \le n], s = 1, \dots, k.$$

Theorem 1.6. *The permanent of the sum of two multiple sampling matrices has the following representation*

$$\text{Per}A_n^m(b_1+a_1,\dots,b_k+a_k)=$$

$$=\sum_{c=0}^{m}\sum_{r_1+\dots+r_k=c}\prod_{s=1}^{k}m_s^{[r_s]}\sum_{q^{(1)}\in Q_{r_1}}\prod_{s=1}^{r_1}a_{1q_s^{(1)}}\sum_{q^{(2)}\in Q_{r_2}^{n-\bar r_2}(q^{(1)})}\prod_{s=1}^{r_2}a_{2q_s^{(2)}}\cdots$$

$$\cdots\sum_{q^{(k)}\in Q_{r_k}^{n-\bar r_k}(q^{(1)},\dots,q^{(k-1)})}\prod_{s=1}^{r_k}a_{kq_s^{(k)}}\text{Per}A_{n-c}^{m-c}(b_1,\dots,b_k;q),\tag{1.40}$$

where

$$A_{n-c}^{m-c}(b_1,\dots,b_k;q)=\begin{bmatrix}A_{n-c}^{m_1-r_1}(b_1;q^{(1)})\\ \dots\\ A_{n-c}^{m_k-r_k}(b_k;q^{(k)})\end{bmatrix},$$

and

$$\bar r_s=c-\sum_{i=s}^{k}r_i,\quad s=1,\dots,k,$$

$$\bar r_1=0,\quad \bar r_k=r_1+\dots+r_{k-1}.$$

The proof of this theorem can be realized by the scheme of the proof of theorem 1.5.

Corollary 1.5. *If for all $s=1,\dots,k$, and $j=1,\dots,n$, $b_{sj}=b_s$, then the decomposition* (1.40) *is reduced to the following*

$$n^{-[m]}\text{Per }A_n^m(b_1+a_1,\dots,b_k+a_k)=$$

$$=\sum_{c=0}^{m}n^{-[c]}\sum_{r_1+\dots+r_k=c}\prod_{s=1}^{k}b_s^{m_s-r_s}m_s^{[r_s]}\times$$

$$\times\text{Per }A_n^c(a_1,\dots,a_k)/\prod_{s=1}^{k}r_s!.\tag{1.41}$$

Corollary 1.6. *If for all $s=1,\dots,k$, and $j=1,\dots,n$, $a_{sj}=a_j$, then the decomposition* (1.40) *is reduced to the following*

$$\text{Per }A_n^m(b_1+a_1,\dots,b_k+a_k)=$$

$$=\sum_{c=0}^{m}\sum_{r_1+\dots+r_k=c}\prod_{s=1}^{k}m_s^{[r_s]}\times$$

$$\times\sum_{1\le j_1\ne\dots\ne j_c\le n}\prod_{s=1}^{c}a_{j_s}\text{Per }A_{n-c}^{m-c}(b_1,\dots,b_k;\ j)/\prod_{s=1}^{k}r_s!.\tag{1.42}$$

Corollary 1.7. *Let $k=2$. Then the decomposition* (1.40) *is reduced to the following formulas:*

$$n^{-[m_1]}\text{Per } A_n^m(1+a_1,a_2) =$$

$$= \sum_{c=0}^{m_1} n^{-[c]} m_1^{[c]} \text{Per } A_n^{c+m_2}(a_1,a_2)/c! \tag{1.43}$$

and

$$n^{-[m_2]}\text{Per } A_n^m(a_1,1+a_2) =$$

$$= \sum_{c=0}^{m_2} n^{-[c]} m_2^{[c]} \text{Per } A_n^{m_1+c}(a_1,a_2)/c!. \tag{1.44}$$

In the general case of two arbitrary matrices

$$A = [a_{ij};\ 1 \le i \le m, 1 \le j \le n]$$

and

$$B = [b_{ij};\ 1 \le i \le m, 1 \le j \le n]$$

the shift decomposition of the sum can be constructed as a generalization of the Laplace formula (1.6).

Introduce the following sets of integer numbers for $1 \le c \le m \le n$

$$Q_{c,m} = \{q = (q_1, \dots, q_c) :\ 1 \le q_1 < \dots < q_c \le m\}.$$

Notice that $Q_{c,n} = Q_c$ by definition in section 1.2.

Theorem 1.7. *For two matrices A and B of $m \times n$ dimension there is the following decomposition*

$$\text{Per } (A+B) = \sum_{c=0}^{m} \sum_{p \in Q_{c,m}} \sum_{q \in Q_c} \text{Per } A[p|q] \text{Per } B(p|q). \tag{1.45}$$

Here for $c = 0$

$$\text{Per } A[p|q] = 1,$$
$$\text{Per } A(p|q) = \text{Per } B.$$

Proof. Use the following identity for arbitrary complex-valued numbers a_s, b_s, $1 \le s \le m$:

$$\prod_{s=1}^{m}(a_s + b_s) = \sum_{c=0}^{m} \sum_{p \in Q_{c,m}} \prod_{s=1}^{c} a_{p_s} \prod_{s=c+1}^{m} b_{p_s} \tag{1.46}$$

where $\{p_1, \dots, p_c, p_{c+1}, \dots, p_m\} = \{1, 2, \dots, m\}$.

By definition of the value of permanent (1.1),

$$\text{Per } (A+B) = \sum_{1 \le j_1 \ne \dots \ne j_m \le n} \prod_{s=1}^{m} (a_{sj_s} + b_{sj_s}).$$

Using (1.46) and changing the order of sums, obtain

$$\text{Per}\ (A+B) = \sum_{c=0}^{m} \sum_{p \in Q_{c,m}} \left\{ \sum_{1 \le j_1 \ne \cdots \ne j_m \le n} \prod_{s=1}^{c} a_{p_s j_{p_s}} \prod_{s=c+1}^{m} b_{p_s j_{p_s}} \right\}. \tag{1.47}$$

The expression in parentheses is the permanent of combined matrices with the $p_1, \ldots, p_c$-th rows of matrix A and $p_{c+1}, \ldots, p_m$-th rows of matrix B.

Therefore, by Laplace formula (1.6), this expression is equal to the inner sum in representation (1.45).

Corollary 1.8. *Let* $B = [b_{ij} = b_i;\ 1 \le i \le m, 1 \le j \le n]$. *Then*

$$\text{Per}\ (A+B) = n^{[m]} \sum_{c=0}^{m} n^{-[c]} \sum_{p \in Q_{c,m}} \prod_{s=c+1}^{m} b_{p_s} \text{Per}\ A[p|q], q \in Q_n. \tag{1.48}$$

Proof. Calculate the expression in parentheses (1.47) to obtain

$$\sum_{1 \le j_1 \ne \cdots \ne j_m \le n} \prod_{s=1}^{c} a_{p_s j_{p_s}} \prod_{s=c+1}^{m} b_{p_s j_{p_s}} =$$

$$= \prod_{s=c+1}^{m} b_{p_s} \sum_{1 \le j_1 \ne \cdots \ne j_m \le n} \prod_{s=1}^{c} a_{p_s j_{p_s}}.$$

Using the decomposition formula (1.11)

$$\sum_{1 \le j_1 \ne \cdots \ne j_m \le n} \prod_{s=1}^{c} a_{p_s j_{p_s}} =$$

$$= \sum_{1 \le j_{p_1} \ne \cdots \ne j_{p_c} \le n} \prod_{s=1}^{c} a_{p_s j_{p_s}} \cdot n^{[m]}/n^{[c]} =$$

$$= n^{[m]} n^{-[c]} \text{Per}\ A[p|q], \quad q \in Q_n.$$

The last two relations, together with (1.47), give (1.48).

Lemma 1.2. *If* $A = [a_{ij};\ 1 \le i \le m, 1 \le j \le n]$ *is a matrix of* $m \times n$, $2 \le m \le n$, *then*

$$\text{Per}\ A = \sum_{\mathcal{D}} \prod_{E \in \mathcal{D}} c(E) r(E), \tag{1.49}$$

where $\mathcal{D}$ consists of all decompositions of the set $\{1, \ldots, m\}$ on non-empty disjoint subsets and

$$c(E) = (-1)^{\mu(E)-1}(\mu(E) - 1)!,$$

where $\mu(E)$ is the number of elements of the set E and

$$r(E) = \sum_{j=1}^{n} \prod_{i \in E} a_{ij}.$$

Proof. (1.49) can be justified by the method of mathematical induction.

Chapter 2

Permanents in the series scheme

Permanents of rectangular matrices with real-valued random elements in the Poisson approximation scheme are considered. The main object under investigation is permanents of multiple functional matrices which can be utilized as statistical images for Ito–Poisson stochastic integrals with a degenerate kernel.

2.1 Poisson approximation scheme

Let $\delta_j^{(n)}$, $1 \leq j \leq n$ be a sample of *random Bernoulli variables* with the distribution

$$\begin{aligned} \mathcal{P}\{\delta_j^{(n)} = 1\} &= 1 - \mathcal{P}\{\delta_j^{(n)} = 0\} \\ &= \lambda/n. \end{aligned} \tag{2.1}$$

It is well-known (see Barbour et al. (1992)) that under condition (2.1) the sums

$$\nu_n := \sum_{j=1}^{n} \delta_j^{(n)}$$

are weakly convergent to the Poisson distributed random variable ν with parameter λ:

$$\nu_n \Rightarrow \nu \text{ as } n \to \infty,$$

$$\mathcal{P}\{\nu = k\} = e^{-\lambda}\lambda^k/k!, \quad k \geq 0. \tag{2.2}$$

Introduce a sampling matrix with random elements

$$A_n^m(\delta^{(n)}) := [a_{ij} = \delta_j^{(n)}; 1 \leq i \leq m, 1 \leq j \leq n]. \tag{2.3}$$

The *random permanent* of this matrix has the value (see (1.2))

$$\begin{aligned} \text{Per } A_n^m(\delta^{(n)}) &= \text{Per} E_{\nu_n}^m \\ &= \nu_n^{[m]} \end{aligned} \tag{2.4}$$

where $E^m_{\nu_n}$ is the matrix with all elements equal to 1 and with dimension $m \times \nu_n$.

Hence, in the Poisson approximation scheme (2.1), the following limiting result (in a weak sense) is true:

$$\mathrm{Per}A^m_n(\delta^{(n)}) \Rightarrow \nu^{[m]} \text{ as } n \to \infty. \tag{2.5}$$

The generating sampling function is defined as follows:

$$\Pi^\delta_n(z) := \prod_{j=1}^n (1 + z\delta^{(n)}_j) \tag{2.6}$$

and can be represented in the form

$$\prod_{j=1}^n (1 + z\delta^{(n)}_j) = (1+z)^{\nu_n}. \tag{2.7}$$

Thus, in the Poisson approximation scheme (2.1), the following result is obtained:

$$\prod_{j=1}^n (1 + z\delta^{(n)}_j) \Rightarrow (1+z)^{\nu} \text{ as } n \to \infty. \tag{2.8}$$

Now consider the random permanent of a sampling matrix with centered random elements

$$A^m_n(\bar\delta^{(n)}) := [a_{ij} = \delta^{(n)}_j - \lambda/n; 1 \le i \le m, 1 \le j \le n]. \tag{2.9}$$

The generating sampling function of centered random variables has the form

$$\begin{aligned} \Pi^{\bar\delta}_n(z) &:= \prod_{j=1}^n (1 + z\bar\delta^{(n)}_j) \\ &= \sum_{m=0}^n \frac{z^m}{m!} \mathrm{Per}A^m_n(\bar\delta^{(n)}_j). \end{aligned} \tag{2.10}$$

Theorem 2.1. *Under the Poisson approximation condition* (2.1) *the weak convergence (in distribution) as* $n \to \infty$

$$\mathrm{Per}A^m_n(\bar\delta^{(n)}_j) \Rightarrow C_m(\nu, \lambda) \tag{2.11}$$

and

$$\begin{aligned} \prod_{j=1}^n (1 + z\bar\delta^{(n)}_j) &\Rightarrow (1+z)^{\nu} e^{-z\lambda} \\ &= \sum_{m=0}^\infty \frac{z^m}{m!} C_m(\nu, \lambda) \end{aligned} \tag{2.12}$$

takes place. The right-hand side of (2.11) *is the stochastic Charlier–Poisson polynomial:*

$$C_m(\nu,\lambda) := \sum_{c=0}^{m}(-\lambda)^{m-c} m^{[c]}\nu^{[c]}/c!. \tag{2.13}$$

Proof. Use the shift decomposition formula (1.32) for permanents and formula (2.4) to obtain

$$\begin{aligned}\mathrm{Per}A_n^m(\bar{\delta}_j^{(n)}) &= \mathrm{Per}A_n^m(-\lambda/n + \delta^{(n)}) \\ &= \sum_{c=0}^{m} q_n(c)(-\lambda)^{m-c} m^{[c]}\nu_n^{[c]}/c!,\end{aligned} \tag{2.14}$$

where

$$\begin{aligned} q_n(c) &:= n^{[m]}/n^{[c]} \cdot n^{m-c} \\ &= (n-c)^{[m-c]}/n^{m-c} \to 1 \end{aligned}$$

as $n \to \infty$.

Take the weak limit in (2.14) to get (2.11).

The generating sampling function (2.10) according to (2.1) has the following representation

$$\prod_{j=1}^{n}(1 + z\bar{\delta}_j^{(n)}) = [1 + z(1-\lambda/n)]^{\nu_n}(1 - z\lambda/n)^{n-\nu_n}.$$

Taking the weak limit and, due to the fact that $\nu_n/n \xrightarrow{P} 0$ as $n \to \infty$, (2.12) is obtained.

The limiting results of theorem 2.1 mean that the permanent of a sampling matrix, with centered Bernoulli distributed random variables under the Poisson approximation scheme (2.1), is the statistical image of the Charlier–Poisson stochastic polynomial. At the same time, the sampling generating function is weak convergent to the generating function of the Charlier–Poisson stochastic polynomials.

2.2 Permanent counting measure

Consider a simple sample in the series scheme

$$\xi_j^{(n)}, \quad 1 \le j \le n \tag{2.15}$$

of independent and identically distributed random variables taking values in a measurable space $(X, \mathcal{X})$ with a distribution function

$$\begin{aligned} P_n(B) &:= \mathcal{P}\{\xi_j^{(n)} \in B\} \\ &= \lambda(B)/n, \quad B \in \mathcal{X}, \end{aligned} \tag{2.16}$$

where $\lambda(B)$ is some Radon measure with $\lambda(X) < \infty$.

Introduce the random δ-measures on $\mathcal{X}$

$$\delta_j^{(n)}(B) := I(\xi_j^{(n)} \in B)$$

$$:= \begin{cases} 1, & \xi_j^{(n)} \in B \\ 0, & \xi_j^{(n)} \notin B. \end{cases} \tag{2.17}$$

The *permanent counting measure* is defined as the permanent of the sampling matrix

$$A_n^m(B) := [a_{ij} = \delta_j^{(n)}(B);\ \ 1 \le i \le m, 1 \le j \le n]. \tag{2.18}$$

So, by analogy to (2.4),

$$\mathrm{Per} A_n^m(B) = \nu_n^{[m]}(B) \tag{2.19}$$

where

$$\nu_n(B) := \sum_{j=1}^{n} \delta_j^{(n)}(B).$$

Under the Poisson approximation condition (2.16), the permanent random measure (2.19) weak converges to the Poisson random measure

$$\mathrm{Per} A_n^m(B) \Rightarrow \nu^{[m]}(B), \quad B \in \mathcal{X} \quad \text{as} \quad n \to \infty, \tag{2.20}$$

where

$$\mathcal{P}\{\nu(B) = k\} = e^{-\lambda(B)} \lambda^k(B)/k!, \quad k \ge 0.$$

The generating sampling function has the following weak limit

$$\prod_{j=1}^{n} [1 + z\delta_j^{(n)}(B)] \Rightarrow (1+z)^{\nu(B)} \quad \text{as} \quad n \to \infty. \tag{2.21}$$

Now consider the random permanent of a sampling matrix with centered random elements

$$A_n^m(B, \lambda) := [a_{ij} = \delta_j^{(n)}(B) - \lambda(B)/n;\ \ 1 \le i \le m, 1 \le j \le n]$$

and the generating sampling function of centered random variables

$$\Pi_n^{\bar{\delta}}(z, B) := \prod_{j=1}^{n} [1 + z\delta_j^{(n)}(B)]$$

$$= \sum_{m=0}^{n} \frac{z^m}{m!} \mathrm{Per} A_n^m(B, \lambda). \tag{2.22}$$

Theorem 2.2. *In the Poisson approximation scheme* (2.16) *the weak convergences as* $n \to \infty$,

$$\mathrm{Per}A_n^m(B,\lambda) \Rightarrow C_m(\nu(B),\lambda(B)) \tag{2.23}$$

and

$$\begin{aligned}\Pi_n^{\bar\delta}(z,B) &\Rightarrow (1+z)^{\nu(B)}e^{-z\lambda(B)}\\ &= \sum_{m=0}^{\infty}\frac{z^m}{m!}C_m(\nu(B),\lambda(B))\end{aligned} \tag{2.24}$$

take place. The right-hand side in (2.23) is the Charlier–Poisson stochastic polynomial

$$C_m(\nu(B),\lambda(B)) := \sum_{c=0}^{m}[-\lambda(B)]^{m-c}m^{[c]}\nu^{[c]}(B)/c!. \tag{2.25}$$

The proof of this theorem is similar to that of theorem 2.1.

Remark 2.1. *It is well-known* (see Ito (1988)) *that the Ito–Poisson stochastic integral for the indicator function $\delta_x(B) = 1$ if $x \in B$ and $\delta_x(B) = 0$ if $x \notin B$ has the following representation*

$$\begin{aligned}I_m^{\nu}(\delta) &:= \int_{X^m}\prod_{s=1}^{m}\delta_{x_s}(B)\nu(dx_s)\\ &= \nu^{[m]}(B)\end{aligned} \tag{2.26}$$

and

$$\begin{aligned}I_m^{\bar\nu}(\delta) &:= \int_{X^m}\prod_{s=1}^{m}\delta_{x_s}(B)\bar\nu(dx_s)\\ &= C_m(\nu(B),\lambda(B)),\end{aligned} \tag{2.27}$$

where

$$\bar\nu(dx) := \nu(dx) - \lambda(dx)$$

is the centered Poisson measure with intensity

$$\lambda(dx) = E\nu(dx).$$

The connection between the Ito–Poisson integrals (2.26) and (2.27) has the following form:

$$I_m^{\bar\nu}(\delta) = \sum_{c=0}^{m}[-\lambda(B)]^{m-c}m^{[c]}I_c^{\nu}(\delta)/c!, \tag{2.28}$$

and

$$I_m^{\nu}(\delta) = \sum_{c=0}^{m}[\lambda(B)]^{m-c}m^{[c]}I_c^{\bar\nu}(\delta)/c!. \tag{2.29}$$

2.3 Permanent multiple counting measure

Let $B_s, 1 \leq s \leq k$ be a collection of disjoint sets from $\mathcal{X}$:

$$B_s \in \mathcal{X}, \ B_s \cap B_{s'} = \emptyset \ \text{if} \ s \neq s'.$$

Introduce the *multiple sampling matrix*

$$A_n^m(B_1, \dots, B_k) := \begin{bmatrix} A_n^{m_1}(B_1) \\ \dots \\ A_n^{m_k}(B_k) \end{bmatrix}, \tag{2.30}$$

where

$$A_n^m(B_s) := [a_{ij} = \delta_j^{(n)}(B_s); \quad 1 \leq i \leq m_s, 1 \leq j \leq n]$$

are the sampling matrices with random δ-measures,

$$E\delta_j^{(n)}(B_s) = \lambda(B_s)/n, \quad 1 \leq s \leq k, \quad m = m_1 + \ldots + m_k.$$

The *permanent multiple counting measure* is defined as Per $A_n^m(B_1, \dots, B_k)$.

Remark 2.2. *In the particular case of $k = m$, the permanent counting measure can be represented with the help of decomposition formula* (1.12) *in the following form:*

$$\text{Per } A_n^m(B_1, \dots, B_m) = \sum_{q_1=1}^{n} \delta_{q_1}^{(n)}(B_1) \times$$

$$\times \sum_{q_2 \in Q_1^{n-1}(q_1)} \delta_{q_2}^{(n)}(B_2) \quad \dots \sum_{q_m \in Q_1^{n-m+1}(q_1, \dots, q_{m-1})} \delta_{q_m}^{(n)}(B_m). \tag{2.31}$$

This formula is valid for any set $B_1 \dots B_m$ and not only for disjoint sets.

Theorem 2.3. *The permanent multiple counting measure has the following representation*

$$\text{Per} A_n^m(B_1, \dots, B_k) = \prod_{s=1}^{k} \nu_n^{[m_s]}(B_s) \tag{2.32}$$

where

$$\nu_n(B_s) := \sum_{j=1}^{n} \delta_j^{(n)}(B_s).$$

In the particular case of $k = m$,

$$\text{Per} A_n^m(B_1, \dots, B_m) = \prod_{s=1}^{m} \nu_n(B_s). \tag{2.33}$$

Proof. Use the decomposition formula (1.11) in the case of $k = 2$, setting $a_{sj} = \delta_j^{(n)}(B_s)$, $s = 1, 2$. The inner sum in (1.11), with factor $m_2!$, is a permanent counting measure

$$m_2! \sum_{q' \in Q_{m_2}^{n-m_1}(q)} \prod_{s=1}^{m_2} \delta_{q'_s}^{(n)}(B_2) = \nu_n^{[m_2]}(B_2) - \sum_{s=1}^{m_1} \delta_{q_s}^{(n)}(B_1).$$

Taking into account that $\delta_j^{(n)}(B_1)\delta_j^{(n)}(B_2) = 0$ it can be verified that

$$\prod_{s=1}^{m_1} \delta_{q_s}^{(n)}(B_1) \sum_{s=1}^{m_1} \delta_{q_s}^{(n)}(B_2) = 0.$$

Hence

$$\begin{aligned}\text{Per } A_n^m(B_1, B_2) &= m_1! \sum_{q \in Q_{m_1}} \prod_{s=1}^{m_1} \delta_{q_s}^{(n)}(B_1)\nu_n^{[m_2]}(B_2) \\ &= \nu_n^{[m_1]}(B_1)\nu_n^{[m_2]}(B_2).\end{aligned}$$

The proof of theorem 2.3, in the general case of $k > 2$, can be realized in a similar way.

Corollary 2.1. *Under the Poisson approximation scheme* (2.16) *the week convergences*

$$\text{Per}A_n^m(B_1, \ldots, B_k) \Rightarrow \prod_{s=1}^{m} \nu^{[m_s]}(B_s), \text{ as } n \to \infty, \tag{2.34}$$

and

$$\prod_{j=1}^{n}[1 + \sum_{s=1}^{k} z_s\delta_j^{(n)}(B_s)] \Rightarrow \prod_{s=1}^{k} (1 + z_s)^{\nu(B_s)}, \text{ as } n \to \infty, \tag{2.35}$$

take place.

Proof. Statement (2.34) is an evident consequence of representation (2.32) and the Poisson approximation condition (2.16), under which $\nu_n(B) \Rightarrow \nu(B)$ as $n \to \infty$. Statement (2.35) follows immediately from the following representation of the generating function

$$\prod_{j=1}^{n}[1 + \sum_{s=1}^{k} z_s\delta_j^{(n)}(B_s)] = \prod_{s=1}^{k}(1 + z_s)^{\nu_n(B_s)}, \tag{2.36}$$

and from the weak convergence of counting measures $\nu_n(B_s) \Rightarrow \nu(B_s)$ as $n \to \infty$.

Remark 2.3. *The limiting result* (2.35) *has the following representation*

$$\prod_{s=1}^{k}(1 + z_s)^{\nu(B_s)} = \exp \int_X \ln[1 + \sum_{s=1}^{k} z_s\delta_x(B_s)]\nu(dx).$$

Now consider the permanent multiple centered counting measure which is determined as the permanent of a multiple sampling matrix

$$A_n^m(B_1,\dots,B_k;\lambda) = \begin{bmatrix} A_n^{m_1}(B_1;\lambda) \\ \dots \\ A_n^{m_k}(B_k;\lambda) \end{bmatrix},$$

where

$$A_n^{m_s}(B_s;\lambda) := [a_{ij} = \delta_j^{(n)}(B_s) - \lambda(B_s)/n;\ \ 1 \le i \le m_s,\ 1 \le j \le n]$$

are sampling matrices with centered random δ-measures and $m = m_1 + \ldots + m_k$.

Theorem 2.4. *Under the Poisson approximation scheme* (2.16) *the weak convergences, as* $n \to \infty$,

$$\text{Per } A_n^m(B_1,\dots,B_k;\lambda) \Rightarrow C_m(\nu(B_1),\dots,\nu(B_k);\lambda) :=$$

$$:= \sum_{c=0}^{m}(-1)^c \sum_{r_1+\ldots+r_k=c} \prod_{s=1}^{k}[\lambda(B_s)]^{m_s-r_s} m_s^{[r_s]} \nu^{[r_s]}(B_s)/r_s! \tag{2.37}$$

and

$$\prod_{j=1}^{n}[1+\sum_{s=1}^{k} z_s(\delta_j^{(n)}(B_s) - \lambda(B_s)/n)] \Rightarrow \prod_{s=1}^{k}(1+z_s)^{\nu(B_s)} e^{-z_s\lambda(B_s)} \tag{2.38}$$

take place.

Proof. Use the shift decomposition formula (1.34), setting $h_s = -\lambda(B_s)/n$, $1 \le s \le k$.

$$\text{Per} A_n^m(B_1,\dots,B_k;\lambda) =$$

$$= \sum_{c=0}^{m} q_n(c) \sum_{r_1+\ldots+r_k=c} \prod_{s=1}^{k}[-\lambda(B_s)]^{m_s-r_s} m_s^{[r_s]} \nu_n^{[r_s]}(B_s)/r_s! \tag{2.39}$$

where

$$q_n(c) := n^c n^{[m]}/n^m n^{[c]} \to 1 \text{ as } n \to \infty.$$

Taking the weak limit in (2.39), set (2.37). The weak limit in (2.38) is due to the following representation of the generating multiple sampling function

$$\prod_{j=1}^{n}[1+\sum_{s=1}^{k} z_s(\delta_j^{(n)}(B_s) - \lambda(B_s)/n)] =$$

$$= [1-\sum_{s=1}^{k} z_s\lambda(B_s)/n]^n \prod_{s=1}^{k}[1+z_s/(1-\sum_{l=1}^{k} z_l\lambda(B_l)/n)]^{\nu_n(B_s)}.$$

It is to be remembered that the sets of $B_s, 1 \le s \le k$, are mutually disjoint and $\delta_j^{(n)}(B_s)\delta_j^{(n)}(B_{s'}) = 0$ as $s \neq s'$.

Remark 2.4. *The limiting result* (2.38) *can be represented as follows*

$$\prod_{s=1}^{k}(1+z_s)^{\nu(B_s)}e^{-z_s\lambda(B_s)} =$$

$$= \exp\left\{\int_X \ln(1+\sum_{s=1}^{k} z_s\delta_x(B_s))\nu(dx) - \sum_{s=1}^{k} z_s\lambda(B_s)\right\}. \tag{2.40}$$

2.4 Permanent functionals

Let $\varphi : X \to R$ be real valued continuous functions with compact support on X. Consider a functional mapping of the simple sample (2.15)

$$\varphi(\xi_j^{(n)}) = \int_X \varphi(x)\delta_j^{(n)}(dx), \quad 1 \le j \le n \tag{2.41}$$

and introduce the sampling functional matrix

$$A_n^m(\varphi) := [a_{ij} = \varphi(\xi_j^{(n)}); \quad 1 \le i \le m, 1 \le j \le n]. \tag{2.42}$$

The *permanent functional* is defined as the permanent of a functional sampling matrix (2.42) and can be represented as the integral over the permanent multiple counting measure (2.41) in the following form:

$$\text{Per } A_n^m(\varphi) = \int_{X^m} \prod_{s=1}^{m} \varphi(x_s)\text{Per } A_n^m(dx_1,\ldots,dx_m). \tag{2.43}$$

By theorem 2.3 obtain

$$\text{Per } A_n^m(\varphi) = \int_{X^m} \prod_{s=1}^{m} \varphi(x_s)\nu_n(dx_s). \tag{2.44}$$

Notice that the integral in (2.44) is not reduced to the product of one-dimensional integrals and can be computed as the Ito–Poisson multiple stochastic integral, taking into account that

$$\nu_n(dx) = \sum_{j=1}^{n} \delta_j^{(n)}(dx)$$

and using

$$\delta_j^{(n)}(dx_s)\delta_j^{(n)}(dx_{s'}) = \begin{cases} \delta_j^{(n)}(dx_s), & s = s', \\ 0, & s \neq s'. \end{cases}$$

For example,

$$\begin{aligned}
\text{Per } A_n^2(\varphi) &= \int_X \int_X \varphi(x_1)\varphi(x_2)\nu_n(dx_1)\nu_n(dx_2) \\
&= \int_X \varphi(x_1)\nu_n(dx_1) \int_X \varphi(x_2)\nu_n(dx_2) \\
&\quad - \int_{X^2} \varphi(x_1)\varphi(x_2) \sum_{j=1}^{n} \delta_j^{(n)}(dx_1)\delta_j^{(n)}(dx_2) \\
&= \left[\int_X \varphi(x)\nu_n(dx)\right]^2 - \int_X \varphi^2(x)\nu_n(dx).
\end{aligned}$$

The generating function of permanent functionals has the following representation

$$\begin{aligned}
\sum_{m=0}^{n} \frac{z^m}{m!} \text{ Per } A_n^m(\varphi) &= \prod_{j=1}^{n}[1 + z\varphi(\xi_j^{(n)})] \\
&= \exp \int_X \ln[1 + z\varphi(x)]\nu_n(dx). \qquad (2.45)
\end{aligned}$$

The permanent functional can be represented as the Charlier–Poisson polynomial of one-dimensional stochastic integrals with respect to the counting measure $\nu_n(dx)$ in the following form:

$$\text{Per } A_n^m(\varphi) = m! \sum_{r_1+2r_2+\cdots+mr_m=m} \prod_{s=1}^{m} (-1)^{r_s-1}[I_1^{\nu_n}(\varphi^s)/s]^{r_s}/r_s!, \qquad (2.46)$$

where

$$I_1^{\nu_n}(\varphi^s) := \int_X \varphi^s(x)\nu_n(dx). \qquad (2.47)$$

Formula (2.46) is obtained by the definition of the terms of the generating function (2.45)

$$\text{Per } A_n^m(\varphi) = \frac{d^m}{dz^m}\left\{\exp \int_X \ln[1 + z\varphi(x)]\nu_n(dx)\right\}\bigg|_{z=0}, \qquad (2.48)$$

and due to the fact that, for any analytical in some neighbourhood of $z = 0$ function $\psi(z)$,

$$\frac{d^m}{dz^m} e^{\psi(z)} = m! e^{\psi(z)} \sum_{r_1+2r_2+\cdots+mr_m=m} \prod_{s=1}^{m} \left[\frac{d^s\psi(z)}{dz^s}\right]^{r_s} / r_s!. \qquad (2.49)$$

Remark 2.5. *Under the Poisson approximation condition* (2.16) *the weak convergences, as* $n \to \infty$,

$$\text{Per } A_n^m(\varphi) \Rightarrow I_m^\nu(\varphi) := \int_{X^m} \prod_{s=1}^{m} \varphi(x_s)\nu(dx_s), \qquad (2.50)$$

and

$$\prod_{j=1}^{n}[1 + z\varphi(\xi_j^{(n)})] \Rightarrow \exp \int_X \ln[1 + z\varphi(x)]\nu(dx) \tag{2.51}$$

take place.

The statements (2.50) and (2.51) hold true taking into account (2.44), (2.45) and due to the fact that, for any finite continuous function φ under condition (2.16),

$$n \int_X \varphi(x) P_n(dx) \to \int_X \varphi(x)\lambda(dx) \quad \text{as } n \to \infty.$$

The right-hand side in (2.50) is the Ito–Poisson multiple stochastic integral with a primitive kernel and the right-hand side in (2.51) is the generating function of such integrals. Therefore the functional permanent can be considered as the *statistical image of the Ito–Poisson integral.*

Similarly, the statistical image of the Ito–Poisson integrals over the centered Poisson measure $\bar{\nu}(dx) = \nu(dx) - \lambda(dx)$ with $\lambda(dx) = E\nu(dx)$ can be constructed as the permanent functional of the sampling matrix

$$A_n^m(\bar{\varphi}) := [a_{ij} = \varphi(\xi_j^{(n)}) - \lambda(\varphi)/n; 1 \le i \le m, 1 \le j \le n], \tag{2.52}$$

where

$$\lambda(\varphi) := \int_X \varphi(x)\lambda(dx).$$

Theorem 2.5. *Under the Poisson approximation condition* (2.16) *the weak convergences, as $n \to \infty$,*

$$\text{Per } A_n^m(\bar{\varphi}) \Rightarrow I_m^{\bar{\nu}}(\varphi) := \int_{X^m} \prod_{s=1}^{m} \varphi(x_s)\bar{\nu}(dx_s) \tag{2.53}$$

and

$$\prod_{j=1}^{n}[1 + z(\varphi(\xi_j^{(n)}) - \lambda(\varphi)/n)] \Rightarrow I^{\bar{\nu}}(z, \varphi)$$

$$:= \exp\left\{\int_X \ln[1 + z\varphi(x)]\nu(dx) - z\lambda(\varphi)\right\} \tag{2.54}$$

take place.

Proof. Similarly to the proof of theorem 2.1, employ the shift decomposition formula (1.34) with $h = -\lambda(\varphi)/n$:

$$\text{Per } A_n^m(\bar{\varphi}) = \text{Per } A_n^m(-\lambda(\varphi)/n + \varphi) =$$

$$= \sum_{c=0}^{m} q_n(c)[-\lambda(\varphi)]^{m-c} m^{[c]} \text{Per } A_n^c(\varphi)/c!\,,$$

where

$$q_n(c) = n^{[m]}n^c/n^{[c]}n^m \to 1$$

as $n \to \infty$.

Taking the weak limit, and due to result (2.50), obtain, as $n \to \infty$,

$$\text{Per } A_n^m(\bar{\varphi}) \Rightarrow \sum_{c=0}^{m}[-\lambda(\varphi)]^{m-c}m^{[c]}I_c^{\nu}(\varphi)/c!. \tag{2.55}$$

But the right-hand side of (2.55) is the Charlier representation of the Ito–Poisson integral over the centered Poisson measure (cf. (2.28)):

$$I_m^{\bar{\nu}}(\varphi) = \sum_{c=0}^{m}[-\lambda(\varphi)]^{m-c}m^{[c]}I_c^{\nu}(\varphi)/c!. \tag{2.56}$$

The generating sampling function in (2.54) is represented as follows

$$\prod_{j=1}^{n}[1 + z(\varphi(\xi_j^{(n)}) - \lambda(\varphi)/n)] = [1 - z\lambda(\varphi)/n]^n \times$$

$$\times \exp \int_X \ln[1 + z\varphi(x)/(1 - z\lambda(\varphi)/n)]\nu_n(dx).$$

Take this weak limit, as $n \to \infty$, to get (2.54).

Remark 2.6. *The right-hand side in (2.54) is the generating function of the Ito–Poisson integrals over the centered Poisson measure:*

$$\exp\left\{\int_X \ln[1 + z\varphi(x)]\nu(dx) - z\lambda(\varphi)\right\} =$$

$$= \sum_{m=0}^{\infty}\frac{z^m}{m!}I_m^{\bar{\nu}}(\varphi) =: I^{\bar{\nu}}(z;\varphi). \tag{2.57}$$

But

$$I^{\bar{\nu}}(z;\varphi) = I^{\nu}(z;\varphi)e^{-z\lambda(\varphi)} \tag{2.58}$$

at the same time.

Here

$$I^{\nu}(z;\varphi) := \sum_{m=0}^{\infty}\frac{z^m}{m!}I_m^{\nu}(\varphi)$$

$$= \exp \int_X \ln[1 + z\varphi(x)]\nu(dx). \tag{2.59}$$

It is easy to check that

$$EI^{\nu}(z;\varphi) = e^{z\lambda(\varphi)}. \tag{2.60}$$

Therefore

$$EI^{\bar{\nu}}(z;\varphi) = 1. \tag{2.61}$$

2.5 Permanent multiple functionals

Consider the multiple Ito–Poisson integral over the Poisson measure $\nu(dx)$ with intensity $\lambda(dx) = E\nu(dx)$ for the degenerate kernel

$$\Phi_k(x_1, \dots, x_m) := \prod_{s=1}^{k} \prod_{r=1}^{m_s} \varphi_s(x_{sr}), \tag{2.62}$$

where $x_{sr} = x_{\bar{m}_s+r}, \bar{m}_s := m_1 + \cdots + m_{s-1}, \bar{m}_1 = 0$,

$$I_m^\nu(\varphi_1, \dots, \varphi_k) := \int_{X^m} \Phi_k(x_1, \dots, x_m) \prod_{s=1}^{m} \nu(dx_s). \tag{2.63}$$

It is well-known (see Ito (1988)) that, for any functions $\varphi_s \in L^2(R; \nu)$, the integral (2.63) is well-defined.

The problem of calculation of the Ito–Poisson integral is not appreciably simple and can be realized with utilization of the generating function

$$I^\nu(z; \varphi_1, \dots, \varphi_k) := \sum_{m=0}^{\infty} \sum_{m_1+\cdots+m_k=m} \prod_{s=1}^{k} \frac{z_s^{m_s}}{m_s!} I_m^\nu(\varphi_1, \dots, \varphi_k)$$

$$= \exp \int_X \ln[1 + \sum_{s=1}^{k} z_s \varphi_s(x)] \nu(dx). \tag{2.64}$$

Analogously, the Ito–Poisson integrals over the centered Poisson measure

$$\bar{\nu}(dx) = \nu(dx) - \lambda(dx)$$

for a degenerate kernel (2.62) are defined as follows

$$I_m^{\bar{\nu}}(\varphi_1, \dots, \varphi_k) = \int_{X^m} \Phi_k(x_1, \dots, x_m) \prod_{s=1}^{m} \bar{\nu}(dx_s). \tag{2.65}$$

The corresponding generating function has the following representation

$$I^{\bar{\nu}}(z; \varphi_1, \dots, \varphi_k) := \sum_{m=0}^{\infty} \sum_{m_1+\cdots+m_k=m} \prod_{s=1}^{k} \frac{z_s^{m_s}}{m_s!} I_m^{\bar{\nu}}(\varphi_1, \dots, \varphi_k)$$

$$= \exp \left\{ \int_X \ln[1 + \sum_{s=1}^{k} z_s \varphi_s(x)] \nu(dx) - \sum_{s=1}^{k} z_s \lambda(\varphi_s) \right\}. \tag{2.66}$$

The connection between functionals (2.63) and (2.65) is determined in the following form:

$$I_m^{\bar{\nu}}(\varphi_1, \dots, \varphi_k) = \sum_{c=0}^{m} (-1)^c \times$$

$$\times \sum_{r_1+\cdots+r_k=c} \prod_{s=1}^{k} [\lambda(\varphi_s)]^{m_s - r_s} m_s^{[r_s]} I_c^\nu(\varphi_1, \dots, \varphi_k)/r_s!. \tag{2.67}$$

The *statistical image* of Ito–Poisson functionals can be constructed as permanents of multiple functional matrices

$$A_n^m(\varphi_1, \dots, \varphi_k) := \begin{bmatrix} A_n^{m_1}(\varphi_1) \\ \dots \\ A_n^{m_k}(\varphi_k) \end{bmatrix},$$

where

$$A_n^{m_s}(\varphi_s) := [a_{ij} = \varphi_s(\xi_j^{(n)});\ 1 \le i \le m_s,\ 1 \le j \le n], \quad 1 \le s \le k,$$

are the sampling matrices defined by a simple sample of random variables $\xi_j^{(n)}$; $1 \le j \le n$, with distributions

$$\begin{aligned} P_n(B) &:= \mathcal{P}\{\xi_j^{(n)} \in B\} \\ &= \lambda(B)/n. \end{aligned} \tag{2.68}$$

Theorem 2.6. *Under the Poisson approximation condition* (2.68) *the weak convergences*

$$\text{Per } A_n^m(\varphi_1, \dots, \varphi_k) \Rightarrow I_m^\nu(\varphi_1, \dots, \varphi_k) \tag{2.69}$$

and

$$\prod_{j=1}^{n}[1 + \sum_{s=1}^{k} z_s\varphi(\xi_j^{(n)})] \Rightarrow I^\nu(z; \varphi_1, \dots, \varphi_k) \tag{2.70}$$

take place, as $n \to \infty$.

Proof. Employ the representation of the permanent multiple functional over the permanent counting measure in the form (see formulas (2.31) and (2.33))

$$\text{Per } A_n^m(\varphi_1, \dots, \varphi_k) = \int_{X_m} \Phi_k(x_1, \dots, x_m) \prod_{s=1}^{m} \nu_n(dx_s) \tag{2.71}$$

and the corresponding generating function in the following form

$$\prod_{j=1}^{n}[1 + \sum_{s=1}^{k} z_s\varphi_s(\xi_j^{(n)})] = \exp \int_X \ln[1 + \sum_{s=1}^{k} z_s\varphi_s(x)]\nu_n(dx). \tag{2.72}$$

Taking weak limits in (2.71) and (2.72) to get the statements of theorem 2.6.

Now consider the multiple functional matrix with centered elements:

$$A_n^m(\bar\varphi_1, \dots, \bar\varphi_k) := \begin{bmatrix} A_n^{m_1}(\bar\varphi_1) \\ \dots \\ A_n^{m_k}(\bar\varphi_k) \end{bmatrix},$$

where

$$A_n^{m_s}(\bar{\varphi}_s) := [a_{ij} = \varphi_s(\xi_j^{(n)}) - \lambda(\varphi_s)/n;\ 1 \le i \le m,\ 1 \le j \le n].$$

Theorem 2.7. *Under the Poisson approximation condition* (2.68) *the weak convergences*

$$\text{Per } A_n^m(\bar{\varphi}_1, \dots, \bar{\varphi}_k) \Rightarrow I_m^{\bar{\nu}}(\varphi_1, \dots, \varphi_k) \tag{2.73}$$

and

$$\prod_{j=1}^{n}[1 + \sum_{s=1}^{k} z_s(\varphi_s(\xi_j^{(n)}) - \lambda(\varphi_s)/n)] \Rightarrow I^{\bar{\nu}}(z; \varphi_1, \dots, \varphi_k) \tag{2.74}$$

take place, as $n \to \infty$.

Proof. Employ the shift decomposition formula (1.34), with $h_s = -\lambda(\varphi_s)/n$. Similarly to the proof of theorem 2.5,

$$\text{Per } A_n^m(\bar{\varphi}_1, \dots, \bar{\varphi}_k) = \sum_{c=0}^{m}(-1)^c q_n(c) \sum_{r_1+\dots+r_k=c} \prod_{s=1}^{k}[\lambda(\varphi_s)]^{m_s - r_s} \times$$

$$\times m_s^{[r_s]} \text{Per } A_n^c(\varphi_1, \dots, \varphi_k)/r_s!, \tag{2.75}$$

where, as usual,

$$q_n(c) = n^c n^{[m]}/n^m n^{[c]} \to 1 \quad \text{as} \quad n \to \infty.$$

Taking the weak limit in (2.75), as $n \to \infty$, and, due to the limiting result (2.69) and representation (2.67), get (2.73).

The limiting result (2.74) is obtained similarly to that in theorem 2.5.

The permanent multiple functionals can be utilized as *statistical images* of Ito–Poisson stochastic integrals with a degenerate kernel (2.62).

Remark 2.7. *A permanent multiple functional can be represented as the integral over the permanent multiple counting measure with a symmetric kernel representable as a permanent* (see remark 2.2):

$$\text{Per } A_n^m(\varphi_1, \dots, \varphi_k) = \frac{1}{m!} \int_{X^m} \text{Per } A_m^m(\varphi_1, \dots, \varphi_k) \prod_{s=1}^{m} \nu_n(dx_s), \tag{2.76}$$

and, analogously,

$$\text{Per } A_n^m(\bar{\varphi}_1, \dots, \bar{\varphi}_k) = \frac{1}{m!} \int_{X^m} \text{Per } A_m^m(\varphi_1, \dots, \varphi_k) \prod_{s=1}^{m} \bar{\nu}_n(dx_s),$$

where

$$\text{Per } A_m^m(\varphi_1, \dots, \varphi_k) := \sum_{\sigma} \prod_{s=1}^{k} \prod_{r=1}^{m_s} \varphi_s(x_{s\sigma(r)}), \tag{2.77}$$

where $\sigma = \{\sigma(1), \ldots, \sigma(m)\}$ *is the set of permutations of numbers* $\{1, 2, \ldots, m\}$.

Therefore the permanent (2.77) is a symmetrized form of the kernel (2.62).

In analogous form the Ito–Poisson integral can be represented:

$$I_m^{\nu}(\varphi_1, \ldots, \varphi_k) = \frac{1}{m!} \int_{X^m} \operatorname{Per} A_m^m(\varphi_1, \ldots, \varphi_k) \prod_{s=1}^{m} \nu(dx_s). \qquad (2.78)$$

2.6 Poisson number of columns

Let N be a Poisson distributed random variable with intensity Λ:

$$\mathcal{P}\{N = n\} = e^{-\Lambda} \Lambda^n / n!, \quad n \geq 0.$$

Consider a sample ξ_j, $1 \leq j \leq N$, of independent identically distributed random variables which take values in a measurable space $(X, \mathcal{X})$ with the distribution

$$P(B) := \mathcal{P}\{\xi_j \in B\}, \quad B \in \mathcal{X}.$$

For a given bounded real-valued function $\varphi : X \to R$, introduce a sampling functional matrix with the Poisson distributed number N of columns:

$$A_N^m(\varphi) := [a_{ij} = \varphi(\xi_j);\ 1 \leq i \leq m,\ 1 \leq j \leq N].$$

Using the counting random measure

$$\nu^{(N)}(B) := \sum_{j=1}^{N} \delta_j(B), \quad \delta_j(B) := I(\xi_j \in B),$$

to obtain the following representations (see Borovskikh et al. (1993a))

$$\operatorname{Per} A_N^m(\varphi) \stackrel{d}{=} \int_X \prod_{s=1}^{m} \varphi(x_s) \nu^{(N)}(dx_s)$$

and

$$\prod_{j=1}^{N} [1 + z\varphi(\xi_j)] = \exp \int_X \ln[1 + z\varphi(x)] \nu^{(N)}(dx).$$

The fact is that the Poisson random measure $\nu_\Lambda(B)$ with intensity

$$\Lambda(B) := E\nu_\Lambda(B) = \Lambda P(B)$$

gives analogous representations of the permanent and generating functions.

Theorem 2.8. *The following representations (in distribution) are true*

$$\operatorname{Per} A_N^m(\varphi) \stackrel{\mathrm{d}}{=} \int_X \prod_{s=1}^{m} \varphi(x_s) \nu_\Lambda(dx_s) = I_m^{\nu_\Lambda}(\varphi)$$

and

$$\prod_{j=1}^{N}[1 + z\varphi(\xi_j)] \stackrel{\mathrm{d}}{=} \exp \int_X \ln[1 + z\varphi(x)]\nu_\Lambda(dx) = I^{\nu_\Lambda}(z;\varphi).$$

Proof. The statement of theorem 2.8 can be verified if it can be proved that the counting random measure $\nu^{(N)}(B)$ is the Poisson measure with intensity $\Lambda(B)$, i.e.

$$\nu^{(N)}(B) \stackrel{\mathrm{d}}{=} \nu_\Lambda(B), \quad B \in \mathcal{X}.$$

Indeed, for any given collection of disjoint subsets B_s, $1 \leq s \leq r$, $\cup_{s=0}^{r} B_s = X$ and fixed non-negative integer numbers n_s, $1 \leq s \leq r$, $\sum_{s=0}^{r} n_s = n$, calculate the conditional distribution of $\nu^{(N)}(B)$ in the following form:

$$\mathcal{P}\{\nu^{(N)}(B_s) = n_s,\ 1 \leq s \leq r | N = n\} = n! \prod_{s=0}^{r} P^{n_s}(B_s)/n_s!.$$

Therefore the unconditional distribution is obtained as follows ($n^0 := \sum_{s=1}^{r} n_s$):

$$\begin{aligned}
\mathcal{P}\{\nu^{(N)}(B_s) &= n_s,\ 1 \leq s \leq r\} = \\
&= e^{-\Lambda} \sum_{n=n^0}^{\infty} \Lambda^n \prod_{s=0}^{r} P^{n_s}(B_s)/n_s! \\
&= e^{-\Lambda} \prod_{s=1}^{r} [\Lambda^{n_s}(B_s)/n_s!] \sum_{n_0=0}^{\infty} \Lambda^{n^0}(B_0)/n^0! \\
&= \prod_{s=1}^{r} [\Lambda^{n_s}(B_s)/n_s!] \exp[-\Lambda + \Lambda(B_0)] \\
&= \prod_{s=1}^{r} [\Lambda^{n_s}(B_s)/n_s!] \exp[-\sum_{s=1}^{r} \Lambda(B_s)] \\
&= \mathcal{P}\{\nu_\Lambda(B_s) = n_s,\ 1 \leq s \leq r\}.
\end{aligned}$$

The proof of theorem 2.8 is completed.

Remark 2.8. *It is easy to verify that*

$$E \prod_{j=1}^{N} [1 + z\varphi(\xi_j)] = \exp\left[- \int_X \varphi(x)\Lambda(dx)\right].$$

Remark 2.9. *Similar results can be obtained for the permanents and the generating function of sampling multiple functional matrices.*

2.7 Stable approximation scheme

Consider the matrix

$$A_n^m(\xi) = [a_{ij} = \xi_j;\ 1 \le i \le m,\ 1 \le j \le n],$$

where $\xi_1, \dots, \xi_n$ are independent identically distributed random variables with $\mu = E\xi_1$. Denote by

$$\eta_{nk}(\xi) = n^{-k/\alpha} \sum_{j=1}^{n} \xi_j^k, \quad k = 1, 2, \dots,$$

where $1 < \alpha < 2$; $\mu = 0$. Assume that the condition of attraction to the stable law is valid, i.e.

$$\eta_{n1}(\xi) \Rightarrow \eta(\alpha), \tag{2.79}$$

where $\eta(\alpha)$ is a random variable which has a stable distribution with basic parameter α. It is well-known that condition (2.79) implies the relation

$$\eta_{nk}(\xi) \Rightarrow \eta(\alpha/k) \quad \text{as} \quad n \to \infty,$$

where $\eta(\alpha/k)$ is a stable random variable with basic parameter α/k (other parameters are defined by k and by the parameters of distribution $\eta(\alpha)$). Thus, if (2.79) is valid and when $n \to \infty$,

$$(\eta_{n1}(\xi), \dots, \eta_{nk}(\xi)) \Rightarrow (\eta(\alpha), \eta(\alpha/2), \dots, \eta(\alpha/k)). \tag{2.80}$$

Here the components of the limit vector are dependent variables.

Convergence of a sampling permanent.
According to (1.5)

$$\text{Per } A_n^m(\xi) = m! P_n^m(\xi),$$

where $P_n^m(\xi)$ is a symmetric polynomial. If $P_n^m(\xi)$ in (1.5) is represented using the Cauchy integral and then this integral is eliminated by the residue Cauchy formula then the second Warring formula can be written

$$P_n^m(\xi/n^{1/\alpha}) = \sum_{i_1+2i_2+\cdots+mi_m=m} (-1)^{m+i_1+\cdots+i_m} \times$$

$$\times \prod_{k=1}^{m} (k^{i_k} i_k!)^{-1} \prod_{k=1}^{m} \eta_{nk}^{i_k}(\xi). \tag{2.81}$$

That is why, by virtue of (2.80),

$$P_n^m(\xi/n^{1/\alpha}) \Rightarrow P_\infty^m(\eta), \tag{2.82}$$

$$P_\infty^m(\eta) = \sum_{i_1+2i_2+\cdots+mi_m=m} (-1)^{i_1+i_2+\cdots+i_m+m} \times$$

$$\times \prod_{k=1}^{m}(k^{i_k}i_k!)^{-1}\prod_{k=1}^{m}\eta^{i_k}(\alpha/k).$$

Consequently for a sampling permanent of fixed degree m under condition (2.79), by virtue of (2.80), the weak convergence

$$n^{-m/\alpha}\mathrm{Per}\ A_n^m(\xi) \Rightarrow m!P_\infty^m(\eta). \tag{2.83}$$

takes place.

Convergence of a multiple sampling permanent.

Consider the matrix

$$A_n^m(\xi_1,\ldots,\xi_k) = \begin{bmatrix} A_n^{m_1}(\xi_1) \\ \vdots \\ A_n^{m_k}(\xi_k) \end{bmatrix},$$

where $\xi_s = (\xi_{sj}, 1 \le j \le n)$, $1 \le s \le k$ are independent random variables with mutually independent components which have identical distribution

$$\mu_s = E\xi_{s1}, \quad s = 1,\ldots,k,$$

when s is fixed.

For $s = 1,\ldots,k$; $p = 1,2,\ldots$ denote

$$\eta_{sp} = n^{-p/\alpha}\sum_{j=1}^{n}\xi_{sj}^p,$$

$$1 < \alpha < 2,$$

and assume that

$$\eta_{s1} \Rightarrow \eta_s(\alpha), \quad s = 1,\ldots,k, \tag{2.84}$$

where $\eta(\alpha)$ is a stable random variable with basic parameter α, and $\eta_1(\alpha),\ldots,\eta_k(\alpha)$ are independent. Analogously, (2.80) can be written as

$$(\eta_{s1},\eta_{s2},\ldots,\eta_{sk}) \Rightarrow (\eta_s(\alpha),\eta_s(\alpha/2),\ldots\eta_s(\alpha/k)), \tag{2.85}$$
$$s = 1,\ldots,k.$$

Theorem 2.9. *If condition* (2.84) *is valid and when* $\mu_s = 0$, $s = 1,\ldots,k$, *and* $m = m_1 + \cdots + m_k$ *is fixed, and* $n \to \infty$, *then*

$$n^{-m/\alpha}\mathrm{Per}\ A_n^m(\xi_1,\ldots,\xi_k) \Rightarrow \prod_{s=1}^{k} m_s!P_\infty^{m_s}(\eta_s^\alpha), \tag{2.86}$$

where

$$\eta_s^\alpha = (\eta_s(\alpha),\eta_s(\alpha/2),\ldots\eta_s(\alpha/k)),$$

and $\eta_1^\alpha, \eta_2^\alpha, \ldots, \eta_m^\alpha$ *are independent.*

Proof. According to (1.12) write

$$n^{-m/\alpha}\text{Per } A_n^m(\xi_1, \ldots, \xi_k) = m_1! n^{-m_1/\alpha} \sum_{q^{(1)} \in Q_{m_1}} \prod_{s=1}^{m_1} \xi_{1q_s^{(1)}} \cdots$$

$$\ldots m_k! n^{-m_k/\alpha} \sum_{q^{(k)} \in Q_{m_k}^{n-\bar{m}_k}(\bar{q}^{(k-1)})} \prod_{s=1}^{m_k} \xi_{kq_s^{(k)}}.$$

To every sum at the right-hand side apply properties (2.81)–(2.82) taking into account that these sums are mutually dependent with respect to indices $q^{(1)}, \ldots, q^{(k-1)}$.

As a result of these actions, and accounting (2.85), arrive at (2.86).

Chapter 3

Permanents with increasing degrees

As in the previous chapter the permanents of sampling matrices are considered in the Poisson approximation scheme, but now their degree m is increasing and satisfying the relation

$$b_n := m/n \to b,$$
$$0 < b < 1, \quad \text{as} \quad n \to \infty. \tag{3.1}$$

3.1 Permanents with shift

Consider a sampling matrix with shift

$$A_n^m(1+\delta^{(n)}) := [a_{ij} = 1 + \delta_j^{(n)}; 1 \le i \le m, 1 \le j \le n], \tag{3.2}$$

constructed by the use of a Bernoulli sample $\delta_j^{(n)}, 1 \le j \le n$ in the series scheme, satisfying the Poisson approximation condition

$$\begin{aligned} \mathcal{P}\{\delta_j^{(n)} = 1\} &= 1 - \mathcal{P}\{\delta_j^{(n)} = 0\} \\ &= \lambda/n \end{aligned} \tag{3.3}$$

with some parameter $\lambda > 0$. Using the centered variables

$$\bar{\delta}_j^{(n)} := \delta_j^{(n)} - \lambda/n \quad \text{with} \quad E\bar{\delta}_j^{(n)} = 0,$$

introduce a sampling matrix with centered shift

$$A_n^m(1+\bar{\delta}^{(n)}) := [a_{ij} = 1 + \bar{\delta}_j^{(n)}; 1 \le i \le m, 1 \le j \le n] \tag{3.4}$$

Theorem 3.1. *In the Poisson approximation scheme (3.3) with increasing degrees (3.1), the weak convergences*

$$n^{-[m]}\mathrm{Per}A_n^m(1+\delta^{(n)}) \Rightarrow (1+b)^\nu \tag{3.5}$$

and

$$n^{-[m]}\mathrm{Per}A_n^m(1+\bar{\delta}^{(n)}) \Rightarrow (1+b)^\nu e^{-\lambda b}\,, \tag{3.6}$$

as $n \to \infty$, take place. So that the permanent of sampling matrices with shift and by increasing degrees can be considered as statistical images of the generating function of the Charlier–Poisson stochastic polynomials .

Proof. The permanent of matrix (3.2) by the shift decomposition formula (1.21) has the representation

$$\begin{aligned} n^{-[m]}\mathrm{Per}A_n^m(1+\delta^{(n)}) &= \sum_{c=0}^{m} m^{[c]}n^{-[c]}\mathrm{Per}A_n^c(\delta^{(n)})/c! \\ &= \sum_{c=0}^{m} m^{[c]}n^{-[c]}\nu_n^{[c]}/c! \end{aligned}$$

by virtue of (2.6). Using the condition (3.1) transform this as follows

$$n^{-[m]}\mathrm{Per}A_n^m(1+\delta^{(n)}) = \sum_{c=0}^{n} b^c\nu_n^{[c]}/c! - \Delta_{n1} - \Delta_{n2}, \tag{3.7}$$

where

$$\begin{aligned} \Delta_{n1} &:= \sum_{c=m+1}^{n} b^c\nu_n^{[c]}/c!, \\ \Delta_{n2} &:= \sum_{c=0}^{m} \varepsilon_n(c)\nu_n^{[c]}/c!, \\ \varepsilon_n(c) &:= b^c - m^{[c]}/n^{[c]}. \end{aligned} \tag{3.8}$$

The first term of (3.7) is the generating function of the permanent of matrices without shift $A_n^c(\delta^{(n)})$, i.e. the generating function of the Bernoulli sample $\delta_j^{(n)}, 1 \le j \le n$, satisfying the Poisson approximation condition (3.3). So that, as in (2.11),

$$\begin{aligned} \sum_{c=0}^{m} b^c\nu_n^{[c]}/c! &= \prod_{j=1}^{n}(1+b\delta_j^{(n)}) \\ &= (1+b)^{\nu_n} \Rightarrow (1+b)^\nu \quad \text{as} \quad n \to \infty. \end{aligned}$$

Now establish the fact that, for increasing degrees (3.1), the second and the third terms of (3.7) satisfy the estimate

$$E|\Delta_{nk}| \to 0 \text{ as } n \to \infty, k = 1, 2. \tag{3.9}$$

Thus the proof of statement (3.5) is complete. First note that considering $\nu_n^{[c]}$ as the permanent of sampling matrix $A_n^c(\delta^{(n)})$ and taking into account (1.2), (2.5) and (2.6),

$$\begin{aligned} E\nu_n^{[c]} &= E\mathrm{Per}A_n^c(\delta^{(n)}) \\ &= n^{[c]}\prod_{j=1}^{c} E\delta_j^{(n)} \\ &= \lambda^c n^{[c]}/n^c \le \lambda^c. \end{aligned}$$

Moreover $E\nu_n^{[c]} \to \lambda^c$, as $n \to \infty$, for fixed c. The first sum in (3.8) allows the following estimate

$$\begin{aligned} E\Delta_{n1} &\le \sum_{c=m+1}^{n} (b\lambda)^c/c! \\ &\le (b\lambda)^{m+1}e^{b\lambda}/(m+1)! \to 0 \quad \text{as} \quad m \to \infty. \end{aligned}$$

Thus, the estimate (3.9) for $k = 1$ is true. Further under condition (3.1),

$$m^{[c]}/n^{[c]} = b_n^c \prod_{s=1}^{c-1}(1 - a_s), \quad a_s \le s/m.$$

Using the algebraic inequality

$$\prod_{s=1}^{c-1}(1 - a_s) > 1 - \sum_{s=1}^{c-1} a_s \quad \text{for} \quad 0 < a_s < 1 \tag{3.10}$$

and, assuming that $m < n$,

$$\begin{aligned} b_n^c &\ge m^{[c]}/n^{[c]} \\ &\ge b_n^c\left(1 - \frac{c(c-1)}{m}\right). \end{aligned}$$

Hence the following estimate holds true, as $m \to \infty$,

$$\begin{aligned} 0 &\le \varepsilon_{n1}(c) := b_n^c - m^{[c]}/n^{[c]} \\ &\le b_n^c c(c-1)/m \to 0. \end{aligned}$$

To complete the estimation of the second sum in (3.8) note that

$$\begin{aligned}\varepsilon_n(c) &:= b^c - m^{[c]}/n^{[c]} \\ &= b^c - b_n^c + \varepsilon_{n1}(c).\end{aligned}$$

Employing the evident inequality

$$|b^c - b_n^c| \le c|b - b_n|$$

conclude that $\varepsilon_n(c)$ satisfies the following estimate

$$|\varepsilon_n(c)| \le c|b - b_n| + c(c-1)b_n^c/m \to 0 \quad \text{as} \quad n, m \to \infty.$$

So, for the second term of (3.8), the estimate (3.10) is obtained as follows:

$$\begin{aligned}E|\Delta_{n2}| &\le \sum_{c=0}^{m} |\varepsilon_n(c)|\lambda^c/c! \\ &\le |b - b_n| \sum_{c=1}^{m} \lambda^c/(c-1)! + \frac{1}{m}\sum_{c=2}^{m}(b_n\lambda)^c/(c-2)! \\ &\le |b - b_n|\lambda e^{\lambda} + (b_n\lambda)^2 e^{b_n\lambda}/m \to 0 \quad \text{as } n, m \to \infty.\end{aligned}$$

Therefore (3.5) holds.
Further, employ the shift decomposition formula (1.32) with the parameter $h = 1 - \lambda/n$

$$\begin{aligned}n^{-[m]}\mathrm{Per}A_n^m(1 + \bar{\delta}^{(n)}) &= \sum_{c=0}^{m}(1 - \lambda/n)^{m-c} m^{[c]} n^{-[c]} \nu_n^{[c]}/c! \\ &= (1 - \lambda/n)^m (1 + b)^{\nu_n} - \\ &- (1 - \lambda/n)^m(\Delta_{n1} + \Delta_{n2}) + \Delta_{n3},\end{aligned} \tag{3.11}$$

where $\Delta_{nk}, k = 1, 2$ are defined in (3.8) and

$$\Delta_{n3} := \sum_{c=0}^{m} q_n(c) m^{[c]} n^{-[c]} \nu_n^{[c]}/c!, \tag{3.12}$$

$$\begin{aligned}q_n(c) &:= (1 - \lambda/n)^{m-c} - (1 - \lambda/n)^m \\ &\le \lambda c/n.\end{aligned} \tag{3.13}$$

Note that it is necessary to estimate the sum of (3.12) using (3.13).

$$E|\Delta_{n3}| \le \sum_{c=1}^{m} \lambda^{c+1}/\{(c-1)!n\}$$
$$\le \lambda^2 e^{\lambda}/n \to 0 \quad \text{as } n \to \infty. \tag{3.14}$$

Taking the limit in (3.11) to get the statement (3.6) of theorem 3.1.

As in section 2.2 consider now the permanent counting measures induced by the permanent of sampling matrices with δ-measure elements.

Let $\xi_j^{(n)}, 1 \le j \le n$ be a sample of random variables taking values in a measurable space $(X, \mathcal{X})$ with a distribution function satisfying the Poisson approximation condition

$$\mathcal{P}_n(B) := \mathcal{P}\{\xi_j^{(n)} \in B\}$$
$$= \lambda(B)/n, \quad B \in \mathcal{X} \tag{3.15}$$

with some bounded Radon measure $\lambda(B)$ $(\lambda(X) < \infty)$. Introduce sampling matrices with shift

$$A_n^m(1 + \delta^{(n)}(B)) := [a_{ij} = 1 + \delta_j^{(n)}(B); 1 \le i \le m, 1 \le j \le n] \tag{3.16}$$

and with centered shift

$$A_n^m(1 + \bar{\delta}^{(n)}(B)) := [a_{ij} = 1 + \bar{\delta}_j^{(n)}(B); 1 \le i \le m, 1 \le j \le n], \tag{3.17}$$

where

$$\delta_j^{(n)}(B) := I(\xi_j^{(n)} \in B)$$
$$:= \begin{cases} 1, & \xi_j^{(n)} \in B, \\ 0, & \xi_j^{(n)} \notin B, \end{cases}$$
$$\bar{\delta}_j^{(n)} := \delta_j^{(n)}(B) - \lambda(B)/n. \tag{3.18}$$

Theorem 3.2. *Under the Poisson approximation condition* (3.15) *and for increasing degrees* (3.1), *the weak convergences*

$$n^{-[m]}\mathrm{Per}A_n^m(1 + \delta^{(n)}(B)) \Rightarrow \exp \int_X \ln[1 + b\delta_x(B)]\nu(dx) \tag{3.19}$$

and

$$n^{-[m]}\mathrm{Per}A_n^m(1 + \bar{\delta}^{(n)}(B)) \Rightarrow \exp\{\int_X \ln[1 + b\delta_x(B)]\nu(dx) - b\lambda(B)\} \tag{3.20}$$

as $n \to \infty$ take place. Here $\nu(B), B \in \mathcal{X}$ is the Poisson random measure with intensity measure $\lambda(B) = E\nu(B)$.

Proof of this theorem is similar to that of theorem 3.1.

3.2 Functionals with shift

Permanent functionals with shift are considered as the permanents of functional sampling matrices with shift:

$$A_n^m(1+\varphi) := [a_{ij} = 1 + \varphi(\xi_j^{(n)}); 1 \le i \le m, 1 \le j \le n] \tag{3.21}$$

and with centered shift

$$A_n^m(1+\bar{\varphi}) := [a_{ij} = 1 + \varphi(\xi_j^{(n)}) - \lambda(\varphi)/n; 1 \le i \le m, 1 \le j \le n] \tag{3.22}$$

where $\varphi(x) : X \to R$ is a real-valued bounded function with compact support on X and a sample $\xi_j^{(n)}, 1 \le j \le n$ satisfies the Poisson approximation condition (3.15), so that

$$E\varphi(\xi_j^{(n)}) = \lambda(\varphi)/n, \tag{3.23}$$

$$\lambda(\varphi) := \int_X \varphi(x)\lambda(dx).$$

Theorem 3.3. *Under the Poisson approximation condition* (3.15) *and for increasing degrees* (3.1) *the weak convergences*

$$n^{-[m]}\mathrm{Per}A_n^m(1+\varphi) \Rightarrow \exp \int_X \ln[1 + b\varphi(x)]\nu(dx) \tag{3.24}$$

and

$$n^{-[m]}\mathrm{Per}A_n^m(1+\bar{\varphi}) \Rightarrow \exp\{\int_X \ln[1 + b\varphi(x)]\nu(dx) - b\lambda(\varphi)\} \tag{3.25}$$

as $n \to \infty$ take place.
So the permanent functionals with shift under the Poisson approximation scheme (3.15) *and with increasing degrees* (3.1) *can be considered as statistical images of generating functions of the Charlier–Poisson functional stochastic polynomials.*

Proof. Acting just as in the proof of theorem 3.1, obtain at first the following representation

$$n^{-[m]}\mathrm{Per}A_n^m(1+\varphi) = \prod_{j=1}^{n}[1 + b\varphi(\xi_j^{(n)})] + \Delta_n\ , \tag{3.26}$$

where Δ_n satisfies the estimate

$$E|\Delta_n| \to 0 \quad \text{as } n \to \infty. \tag{3.27}$$

Setting

$$\nu_n(B) := \sum_{j=1}^{n} \delta_j^{(n)}(B)$$

to get, for the first term of (3.26), the following representation

$$\prod_{j=1}^{n}[1 + b\varphi(\xi_j^{(n)})] = \exp \int_X \ln[1 + b\varphi(x)]\nu_n(dx) \ . \tag{3.28}$$

Under the Poisson approximation condition (3.15), $\nu_n(B) \Rightarrow \nu(B)$ as $n \to \infty$ with $E\nu(B) = \lambda(B)$. Taking the limit in (3.23) and by virtue of (3.27) and (3.28), obtain (3.24).

The proof of (3.25) is the consequence of the following representation

$$n^{-[m]}\text{Per}A_n^m(1 + \bar{\varphi}) =$$

$$= (1 - \lambda(\varphi)/n)^m \prod_{j=1}^{n}[1 + b\varphi(\xi_j^{(n)})] + \Delta_n \ , \tag{3.29}$$

where Δ_n satisfies the estimate (3.27). Notice that for increasing degrees (3.1)

$$(1 - \lambda(\varphi)/n)^m \to e^{-b\lambda(\varphi)} \ .$$

Taking the limit in (3.29) to obtain (3.25).

3.3 Permanent multiple functionals with shift

At first study the permanent of a two-sampling functional matrix with shift.

$$A_n^m(1 + \varphi_1,\ 1 + \varphi_2) := \begin{bmatrix} A_n^{m_1}(1 + \varphi_1) \\ \\ A_n^{m_2}(1 + \varphi_2) \end{bmatrix}, \tag{3.30}$$

where $A_n^{m_s}(1 + \varphi_s)$, $s = 1, 2$ are sampling matrices with elements

$$a_{ij}^{(s)} := 1 + \varphi_s(\xi_j^{(n)}), \quad 1 \le i \le m_s, \ 1 \le j \le n, \ s = 1, 2.$$

Theorem 3.4. *Assume that the Poisson approximation scheme* (3.3) *holds true and the degrees* m_s, $s = 1, 2$ *are increasing in the following way*

$$m_s/n \longrightarrow b_s,$$

$$0 < b_s < 1, \quad s = 1, 2, \ n \to \infty. \tag{3.31}$$

Then the weak convergence

$$n^{-[m]}\mathrm{Per}A_n^m(1+\varphi_1, 1+\varphi_2) \Longrightarrow \exp(\int_X \ln[1+b_1\varphi_1(x)+b_2\varphi_2(x)]\nu(dx))$$

takes place.

Proof. Use the decomposition formula (1.11)

$$n^{-[m]}\mathrm{Per}A_n^m(1+\varphi_1, 1+\varphi_2) =$$

$$= \frac{m_1!m_2!}{n^{[m]}} \sum_{q\in Q_{m_1}} \prod_{s=1}^{m_1}(1+\zeta_{1q_s}^{(n)}) \sum_{q\in Q_{m_2}^{n-m_1}(q)} \prod_{s=1}^{m_2}(1+\zeta_{2q'_s}^{(n)}), \tag{3.32}$$

where

$$\zeta_{sj}^{(n)} := \varphi_s(\xi_j^{(n)}), 1 \le j \le n, \; s = 1,2.$$

The inner sums in (3.32) with factor $m_2!$ are permanents of sampling matrices $A_{n-m_1}^{m_2}(1+\zeta_2; q)$ with elements

$$a_{ij} = 1+\zeta_{2j}^{(n)}, \; 1\le j\le n, \; j \ne q_s, \; 1 \le s \le m_1.$$

By the shift decomposition formula (1.22), as in the proof of theorem 3.2,

$$(n-m_1)^{-[m_2]}\mathrm{Per}A_{n-m_1}^{m_2}(1+\zeta_2; q)$$

$$= \sum_{c=0}^{m_2} \frac{m_2^{[c]}}{(n-m_1)^{[c]}c!}\mathrm{Per}A_{n-m_1}^{c}(\zeta_2; q). \tag{3.33}$$

The sum in (3.33) can be transformed as follows

$$\sum_{c=0}^{n-m_1} \frac{m_2^{[c]}}{(n-m_1)^{[c]}c!}\mathrm{Per}A_{n-m_1}^{c}(\zeta_2; q)$$

$$= \sum_{c=0}^{n-m_1} b^c\mathrm{Per}A_{n-m_1}^{c}(\zeta_2; q)/c! + \Delta_n(q) \,, \tag{3.34}$$

where $b := b_2/(1-b_1)$ and

$$\Delta_n(q) := \sum_{c=0}^{m_2}\left(\frac{m_2^{[c]}}{(n-m_1)^{[c]}} - b^c\right)\mathrm{Per}A_{n-m_1}^{c}(\zeta_2; q)/c! +$$

$$+ \sum_{c=m_2+1}^{n-m_1} b^c\mathrm{Per}A_{n-m_1}^{c}(\zeta_2; q)/c! \,. \tag{3.35}$$

Note that $\Delta_n(q)$ satisfies the condition (3.26). The sum in (3.35) is the generating function of the permanents

$$\sum_{c=0}^{n-m_1} b^c \mathrm{Per} A^c_{n-m_1}(\zeta_2; q)/c!$$

$$= \prod_{j=1}^{n}[1 + b\zeta_{2j}^{(n)}] \prod_{s=1}^{m_1}[1 + b\zeta_{2q_s}^{(n)}]^{-1}. \tag{3.36}$$

Insert (3.33) - (3.36) into (3.32) to get

$$n^{-[m]}\mathrm{Per}A^m_n(1+\varphi_1, 1+\varphi_2) = \frac{m_1!(n-m_1)^{[m_2]}}{n^{[m]}} \times$$

$$\times \sum_{q \in Q_{m_1}} [1 + \zeta_{1q_s}^{(n)}][1 + b\zeta_{2q_s}^{(n)}]^{-1} \prod_{j=1}^{n}[1 + b\zeta_{2j}^{(n)}] + \Delta_n , \tag{3.37}$$

where

$$\Delta_n = \frac{m_1!(n-m_1)^{[m_2]}}{n^{[m]}} \sum_{q \in Q_{m_1}} \prod_{s=1}^{m_1}(1 + \zeta_{1q_s}^{(n)})\Delta_n(q). \tag{3.38}$$

Comparing (3.36) with this relation, conclude that $\prod_{s=1}^{m_1}(1 + \zeta_{1q_s}^{(n)})$ and $\Delta_n(q)$ are independent random variables. Hence

$$E|\Delta_n| \le \frac{m_1!(n-m_1)^{[m_2]}}{n^{[m]}} \sum_{q \in Q_{m_1}} (1 + E|\zeta_{11}^{(n)}|)^{m_1} E|\Delta_n(q)| . \tag{3.39}$$

Here

$$E|\zeta_{11}^{(n)}| = \int_X |\varphi_1(x)| P_n(dx)$$

$$= \int_X |\varphi_1(x)| \lambda(dx)/n . \tag{3.40}$$

Hence, under condition (3.32),

$$(1 + E|\zeta_{11}^{(n)}|)^{m_1} \to e^{\lambda_1 b_1},$$

$$\lambda_1 := \int_X |\varphi_1(x)|\lambda(dx)$$

$$= \lambda(|\varphi_1|). \tag{3.41}$$

Thus, from (3.40) - (3.41), it can be concluded that Δ_n satisfies the condition (3.27).

The sum in (3.37) is transformed as follows:

$$m_1! \sum_{q \in Q_{m_1}} \prod_{s=1}^{m_1} [1 + \zeta_{1q_s}^{(n)}][1 + b\zeta_{2q_s}^{(n)}]^{-1} =$$

$$= m_1! \sum_{q \in Q_{m_1}} \prod_{s=1}^{m_1} [1 + \eta_{1q_s}^{(n)}] =:$$

$$=: \mathrm{Per} A_n^{m_1}(1 + \eta) , \tag{3.42}$$

where

$$\eta_j^{(n)} := (\zeta_{1j}^{(n)} - b\zeta_{2j}^{(n)})(1 + \zeta_{2j}^{(n)})^{-1} . \tag{3.43}$$

As in theorem 3.2, for the permanent functional with shift , the following asymptotic representation (see (3.29)) is obtained

$$n^{-[m]} \mathrm{Per} A_n^m(1 + \eta) =$$

$$= \prod_{j=1}^{n} (1 + b_1 \eta_j^{(n)}) + \Delta'_n \tag{3.44}$$

with Δ'_n satisfying (3.27).

Further, using (3.44) and taking into account that $b(1 - b_1) = b_2$,

$$\prod_{j=1}^{n} (1 + b_1 \eta_j^{(n)}) =$$

$$= \prod_{j=1}^{n} [1 + b_1 \zeta_{1j}^{(n)} + b_2 \zeta_{2j}^{(n)}] \prod_{j=1}^{n} (1 + b\zeta_{2j}^{(n)})^{-1}. \tag{3.45}$$

Combine (3.42) - (3.45) with (3.37) to obtain

$$n^{-[m]} \mathrm{Per} A_n^m(1 + \varphi_1, 1 + \varphi_2)$$

$$= \frac{n^{[m_1]}(n - m_1)^{[m_2]}}{n^{[m]}} \prod_{j=1}^{n} [1 + b_1 \zeta_{1j}^{(n)} + b_2 \zeta_{2j}^{(n)}] + \Delta_n$$

$$= \exp \int_X \ln(1 + b_1 \varphi_1(x) + b_2 \varphi_2(x)) \nu_n(dx) + \Delta_n .$$

Note that

$$n^{[m_1]}(n - m_1)^{[m_2]} / n^{[m]} = 1.$$

The last asymptotic representation completes the proof of theorem 3.4. In the general case $k \geq 2$, the following theorem is valid.

Theorem 3.5. *Under the Poisson approximation condition* (3.15) *and for increasing degrees*

$$b_{ns} := m_s/n \longrightarrow b_s \text{ as } n \to \infty,$$

$$0 < b_s < 1, \quad 1 \le s \le k \tag{3.46}$$

the weak convergences

$$n^{-[m]}\mathrm{Per}A_n^m(1+\varphi_1,\dots,1+\varphi_k) \Rightarrow \exp\int_X \ln[1+\sum_{s=1}^k b_s\varphi_s(x)]\nu(dx) \tag{3.47}$$

and

$$n^{-[m]}\mathrm{Per}A_n^m(1+\bar\varphi_1,\dots,1+\bar\varphi_k) \Rightarrow$$

$$\Rightarrow \exp\left[\int_X \ln[1+\sum_{s=1}^k b_s\varphi_s(x)]\nu(dx) - \sum_{s=1}^k b_s\lambda(\varphi_s)\right] \tag{3.48}$$

take place.

Proof. Employ the shift decomposition formula (1.33)

$$n^{-[m]}\mathrm{Per}A_n^m(1+\varphi_1,\dots,1+\varphi_k)$$

$$= \sum_{c=0}^m n^{-[c]} \sum_{r_1+\dots+r_k=c} \prod_{s=1}^k m_s^{[r_s]}\mathrm{Per}A_n^c(\varphi_1,\dots,\varphi_k)/r_s!\,.$$

Substituting nb_{ns} instead of m_s to obtain

$$n^{-[m]}\mathrm{Per}A_n^m(1+\varphi_1,\dots,1+\varphi_k) =$$

$$= \sum_{c=0}^n \sum_{r_1+\dots+r_k=c} \prod_{s=1}^k b_{ns}^{r_s}\mathrm{Per}A_n^c(\varphi_1,\dots,\varphi_k)/r_s! - \Delta_{n1} - \Delta_{n2} \tag{3.49}$$

where

$$\Delta_{n1} := \sum_{c=m+1}^n \sum_{r_1+\dots+r_k=c} \prod_{s=1}^k b_{ns}^{r_s}\mathrm{Per}A_n^c(\varphi_1,\dots,\varphi_k)/r_s!\,, \tag{3.50}$$

$$\Delta_{n2} := \sum_{c=0}^m \sum_{r_1+\dots+r_k=c} q_n^c(r)\prod_{s=1}^k \mathrm{Per}A_n^c(\varphi_1,\dots,\varphi_k)/r_s!\,, \tag{3.51}$$

$$q_n^c(r) := \prod_{s=1}^k b_{ns}^{r_s} - \prod_{s=1}^k m_s^{[r_s]}/n^{[c]}. \tag{3.52}$$

The first term in (3.49) is the generating function of multiple sampling permanents

$$\sum_{c=0}^{n} \sum_{r_1+\ldots+r_k=c} \prod_{s=1}^{k} b_{ns}^{r_s} \mathrm{Per} A_n^c(\varphi_1, \ldots, \varphi_k)/r_s!$$

$$= \prod_{j=1}^{n} [1 + \sum_{s=1}^{k} b_{ns}\varphi_s(\xi_j^{(n)})]$$

$$= \exp \int_X \ln[1 + \sum_{s=1}^{k} b_{ns}\varphi_s(x)]\nu_n(dx)$$

$$\Rightarrow \exp \int_X \ln[1 + \sum_{s=1}^{k} b_s\varphi_s(x)]\nu(dx) \quad \text{as} \quad n \to \infty$$

under condition of this theorem.

To complete the proof of theorem 3.5, verify the estimates

$$E|\Delta_{nk}| \longrightarrow 0 \quad \text{as} \quad n \to \infty, k = 1, 2. \tag{3.53}$$

Note that, by the definition of the permanent,

$$E|\mathrm{Per} A_n^c(\varphi_1, \ldots, \varphi_k)| \le n^{[c]} \prod_{s=1}^{k} \lambda^{r_s}(|\varphi_s|)/n^c$$

$$\le \prod_{s=1}^{k} \lambda^{r_s}(|\varphi_s|).$$

Therefore

$$E|\Delta_{n1}| \le \sum_{c=m+1}^{n} \sum_{r_1+\ldots+r_k=c} \prod_{s=1}^{k} \lambda^{r_s}(|\varphi_s|)/r_s!$$

$$= \sum_{c=m+1}^{n} [\sum_{s=1}^{k} \lambda(|\varphi_s|)]^c/c!$$

$$\le [\sum_{s=1}^{k} \lambda(|\varphi_s|)]^{m+1} \prod_{s=1}^{k} e^{\lambda(|\varphi_s|)}/(m+1)! \longrightarrow 0 \quad \text{as} \quad m \to \infty.$$

Note also that, in the sum (3.51), the first two terms with $c = 0, 1$ are equal to 0, because $r_1 + \ldots + r_k = 0$ yields $r_1 = \ldots = r_k = 0$ and the equation $r_1 + \ldots + r_k = 1$ has k solutions $(1, 0, \ldots, 0)$, $\ldots$, $(0, \ldots, 0, 1)$, so that $q_n^c(r) = 0$ as $c = 0, 1$.
Using the condition (3.46) it can be verified that, for $c \ge 2$,

$$q_n^c(r) = \prod_{s=1}^{k} (m_s/n)^{r_s} - \prod_{s=1}^{k} m_s^{[r_s]}/n^{[c]}$$

$$\le q_n c(c-1),$$

where

$$\begin{aligned} q_n &:= n^{-1} + \sum_{s=1}^{k} m_s^{-1} \\ &= (1 + \sum_{s=1}^{k} b_{ns}^{-1})/n \to 0 \quad \text{as} \quad n \to \infty. \end{aligned}$$

The third term of (3.49) satisfies the estimate

$$\begin{aligned} E|\Delta_{n2}| &\leq q_n \sum_{c=2}^{m} [\sum_{s=1}^{k} \lambda(|\varphi_s|)]^c/(c-2)! \\ &\leq (k+1)(1 + \sum_{s=1}^{k} b_s^{-1})[\sum_{s=1}^{k} \lambda(|\varphi_s|)]^2 \\ &\quad \times \prod_{s=1}^{k} e^{\lambda(|\varphi_s|)}/n \to 0 \quad \text{as} \quad n \to \infty. \end{aligned}$$

To finish the proof of (3.48) employ the shift decomposition formula (1.41) with parameters $h_s = 1 - \lambda(\varphi_s)/n, 1 \leq s \leq k$,

$$n^{-[m]}\mathrm{Per}A_n^m(1+\bar{\varphi}_1, \ldots, 1+\bar{\varphi}_k) =$$

$$= \sum_{c=0}^{m} n^{-[c]} \sum_{r_1+\ldots+r_k=c} \prod_{s=1}^{k} h_s^{m_s - r_s} m^{[r_s]} \mathrm{Per}A_n^c(\varphi_1, \ldots, \varphi_k)/r_s!.$$

Now, similarly to (3.49), transform it to the following representation

$$n^{-[m]}\mathrm{Per}A_n^m(1+\bar{\varphi}_1, \ldots, 1+\bar{\varphi}_k) = \prod_{s=1}^{k}(1 - \lambda(\varphi_s)/n)^{m_s} \times$$

$$\times \left[\prod_{j=1}^{n} [1 + \sum_{s=1}^{k} b_{ns}\varphi_s(\xi_j^{(n)})] - \Delta_{n1} - \Delta_{n2} \right] + \Delta_{n3}\,, \tag{3.54}$$

where $\Delta_{nk}, k = 1, 2$ are defined in (3.51) - (3.53) and

$$\Delta_{n3} := \sum_{c=1}^{m} n^{-[c]} \sum_{r_1+\ldots+r_k=c} q_n^c(r,h) \prod_{s=1}^{k} m_s^{[r_s]} \mathrm{Per}A_n^c(\varphi_1, \ldots, \varphi_k)/r_s!,$$

where

$$\begin{aligned} q_n^c(r,h) &:= \prod_{s=1}^{k} h_s^{m_s - r_s} - \prod_{s=1}^{k} h_s^{m_s} \\ &= \prod_{s=1}^{k} h_s^{m_s - r_s}(1 - \prod_{s=1}^{k} h_s^{r_s}). \end{aligned}$$

As $n \to \infty$

$$0 < \lambda(|\varphi_s|)/n < 1,$$

$$0 < h_s < 1, \quad 1 \le s \le k.$$

Then

$$\begin{aligned} |q_n^c(r,h)| &\le 1 - \prod_{s=1}^k h_s^{r_s} \\ &\le \sum_{s=1}^k r_s |\lambda(\varphi_s)|/n \\ &\le c \sum_{s=1}^k \lambda(|\varphi_s|)/n. \end{aligned}$$

Therefore the following estimate is obtained:

$$\begin{aligned} E|\Delta_{n3}| &\le \frac{1}{n} \sum_{s=1}^k \lambda(|\varphi_s|) \sum_{c=1}^m [\sum_{s=1}^k \lambda(|\varphi_s|)]^c/(c-1)! \\ &\le \frac{1}{n} [\sum_{s=1}^k \lambda(|\varphi_s|)]^2 \prod_{s=1}^k e^{\lambda(|\varphi_s|)} \to 0 \quad \text{as } n \to \infty \,. \end{aligned}$$

To complete the proof of theorem 3.5 note that the first term of (3.54) converges weakly, as $n \to \infty$,

$$\prod_{s=1}^k (1 - \lambda(\varphi_s)/n)^{m_s} \prod_{j=1}^n [1 + \sum_{s=1}^k b_{ns} \varphi_s(\xi_j^{(n)})]$$

$$= \prod_{s=1}^k (1 - \lambda(\varphi_s)/n)^{nb_{ns}} \exp \int_X \ln[1 + \sum_{s=1}^k b_{ns} \varphi_s(x)] \nu_n(dx)$$

$$\Rightarrow \exp \left\{ \int_X \ln[1 + \sum_{s=1}^k b_s \varphi_s(x)] \nu(dx) - \sum_{s=1}^k b_s \lambda(\varphi_s) \right\} .$$

Under the conditions of theorem 3.5 the permanent multiple functionals can be considered as statistical images of generating functions of the Charlier–Poisson functionals with degenerate kernels.

3.4 Stable approximation scheme

Consider a random permanent Per $A_n^m(\xi)$ under the conditions of section 2.7, but suppose that $\mu \neq 0$. Therefore the condition (2.79) of attraction to stable law means that

$$\eta_{n1}(\xi - \mu) \Rightarrow \eta(\alpha). \tag{3.55}$$

Theorem 3.6. *If $m \to \infty$ and $n \to \infty$ and*

$$mn^{(1-\alpha)/\alpha} \to \gamma, \; \gamma > 0, \tag{3.56}$$

then, under condition (3.55), *the weak convergence*

$$n^{-[m]}\text{Per } A_n^m(\xi/\mu) \Rightarrow \sum_{c=0}^{\infty}(\gamma/\mu)^c P_\infty^c(\eta), \tag{3.57}$$

takes place. $P_\infty^c(\eta)$ is defined in (2.82) *when $m = c$.*

Proof. According to formula (1.32) write

$$n^{-[m]}\mu^{-m}\text{Per } A_n^m(\xi) =$$

$$= \sum_{c=0}^{m} n^{c/\alpha} m^{[c]} n^{-[c]} \mu^{-c} P_n^c((\xi - \mu)/n^{1/\alpha}).$$

Here, under condition (3.55),

$$P_n^c((\xi - \mu)/n^{1/\alpha}) \Rightarrow P_\infty^c(\eta),$$

and, under condition (3.56),

$$n^{c/\alpha} m^{[c]} n^{-[c]} \to \gamma^c,$$

that leads, as a result, to (3.57).

Chapter 4

Normal approximation

Natural asymptotic analysis of generating sampling functions and permanents is considered in the normal approximation scheme for sums of independent random variables. Here interesting statistical interpretation of limiting results can be expected.

4.1 Bernoulli scheme

Let ξ_j, $1 \leq j \leq n$, be a sample of Bernoulli random variables with probabilities

$$\begin{aligned} p &= \mathcal{P}\{\xi_j = a\} \\ &= 1 - \mathcal{P}\{\xi_j = -b\} \\ &= 1 - q. \end{aligned}$$

Here $a > 0, \quad b > 0, \quad a + b = c, \quad \rho = b/c.$

The asymptotic behaviour as $n \to \infty$ of the permanent of a sampling matrix

$$A_n^m(\xi) := [a_{ij} = \xi_j; 1 \leq i \leq m, 1 \leq j \leq n]$$

and of the generating sampling function

$$\prod_{j=1}^{n}(1 + z\xi_j) = \sum_{m=0}^{n} \frac{z^m}{m!} \mathrm{Per} A_n^m(\xi) \tag{4.1}$$

is of prime interest.

Introduce the following notations

$$\begin{aligned} \mu &= E\xi_j \\ &= ap - bq \\ &= c(p - \rho), \end{aligned}$$

$$\sigma^2 = E(\xi_j - \mu)^2$$

$$\begin{aligned} &= a^2p + b^2q - \mu^2 \\ &= c^2pq, \end{aligned}$$

$$\begin{aligned} v &= \mu/\sigma \\ &= (p-\rho)/\sqrt{pq}\,. \end{aligned} \tag{4.2}$$

The *accumulated frequencies* of sampling values are denoted as follows:

$$\begin{aligned} \nu_{na} &:= \sum_{j=1}^{n} I(\xi_j = a), \\ \nu_{nb} &:= \sum_{j=1}^{n} I(\xi_j = -b). \end{aligned} \tag{4.3}$$

Note that $\nu_{na} + \nu_{nb} = n$.

Proposition 4.1. *The generating sampling function and the permanent of a sampling matrix have the following frequency representation:*

$$\prod_{j=1}^{n}(1 + z\xi_j) = (1+za)^{\nu_{na}}(1-zb)^{\nu_{nb}} \tag{4.4}$$

and

$$\mathrm{Per}A_n^m(\xi) = \sum_{r=0}^{m} a^r(-b)^{m-r}\nu_{na}^{[r]}\,\nu_{nb}^{[m-r]}\;m^{[r]}/r!. \tag{4.5}$$

Proof. The first statement (4.4) follows from the fact that, for the two-valued random variable ξ, the relation

$$1 + z\xi = (1+za)^{I(\xi=a)}(1-zb)^{I(\xi=-b)}$$

holds.

The frequency representation for the permanent follows from the property that the generating sampling function is the product of two generating functions

$$\begin{aligned} (1+za)^{\nu_{na}} &= \sum_{m=0}^{\nu_{na}} \frac{z^m}{m!}a^m\nu_{na}^{[m]}, \\ (1-zb)^{\nu_{nb}} &= \sum_{m=0}^{\nu_{nb}} \frac{z^m}{m!}(-b)^m\nu_{nb}^{[m]}. \end{aligned}$$

So the frequency representation (4.5) is the convolution of two sequences

$$\{a^m\nu_{na}^{[m]}\}$$

and

$$\{(-b)^m \nu_{nb}^{[m]}\}.$$

Remark 4.1. *The right-hand side in* (4.5) *is the stochastic Kravchuk's polynomial of a binomially distributed random variable* ν_{na} (or ν_{nb}) (see Korolyuk (1993)). *By analogy, the stochastic Kravchuk's polynomials of multinomially distributed random variables as the permanent of a sampling matrix defined by a sample of random variables can be constructed, which take a finite number of values* (see Accardi et al. (1993)).

Using the frequency representation (4.4) and (4.5) the asymptotic behavior as $n \to \infty$ for a fixed degree m, as well as the generating sampling function, can be investigated as the permanent.

Proposition 4.2. *The normalized generating sampling function of the Bernoulli sample is weakly convergent as* $n \to \infty$, *namely*

$$\prod_{j=1}^{n}(1 + z\xi_j/\sigma\sqrt{n}) \exp[-zv\sqrt{n}] \Rightarrow \exp[z\tau - z^2(1+v^2)/2], \tag{4.6}$$

where τ *is the standard normally distributed random variable with* $E\tau = 0,\ E\tau^2 = 1$.

Proof. It is easy to verify that the frequencies (4.3) have the following representation

$$\nu_{na} = n\rho + \tau_n/c,$$

$$\nu_{nb} = n(1-\rho) + \tau_n/c,$$

where

$$\tau_n := \sum_{j=1}^{n} \xi_j.$$

Taking a logarithm in the frequency representation (4.4) and using the Taylor formula for a small z, obtain the following asymptotic representation as $n \to \infty$:

$$\begin{aligned} \ln\prod_{j=1}^{n}(1 \ &+ \ z\xi_j/\sigma\sqrt{n}) = \\ &= \ \nu_{na}\ln(1 + za/\sigma\sqrt{n}) + \nu_{nb}\ln(1 - zb/\sigma\sqrt{n}) \\ &= \ zv\sqrt{n} + z\tau_{n1} - z^2\tau_{n2}/2 + \Delta_n, \end{aligned} \tag{4.7}$$

where

$$\tau_{n1} := \sum_{j=1}^{n}(\xi_j - \mu)/\sigma\sqrt{n} \Rightarrow \tau\ , \quad \text{as } n \to \infty \tag{4.8}$$

by the central limit theorem (Loeve (1960)) and

$$\begin{aligned}\tau_{n2} &:= [a^2\nu_{na} + b^2\nu_{nb}]/n\sigma^2 \\ &= \sum_{j=1}^{n} \xi_j^2/n\sigma^2 \\ &\xrightarrow{P} 1 + v^2 \end{aligned} \tag{4.9}$$

by the law of large numbers (Loeve (1960)). Here

$$\Delta_n \xrightarrow{P} 0 \quad \text{as} \quad n \to \infty.$$

Note that, according to the notation (4.3),

$$\begin{aligned}\nu_{na}/n &\xrightarrow{P} p, \\ \nu_{nb}/n &\xrightarrow{P} q, \quad \text{as} \quad n \to \infty.\end{aligned}$$

So the asymptotic representation (4.7) can be rewritten as follows:

$$\prod_{j=1}^{n}(1 + z\xi_j/\sigma\sqrt{n})\exp[-zv\sqrt{n}] = \exp[z\tau_{n1} - z^2\tau_{n2}/2] + \Delta_n. \tag{4.10}$$

Taking into account (4.8) and (4.9) to get the weak limit in (4.10) as $n \to \infty$ in the form (4.6).

Remark 4.2. *The main part of the right-hand side of* (4.10) *is the generating function of the Hermite polynomials of two variables:*

$$\exp[z\tau_{n1} - z^2\tau_{n2}/2] = \sum_{m=0}^{\infty} \frac{z^m}{m!} H_m(\tau_{n1}, \tau_{n2}). \tag{4.11}$$

Comparing (4.10) *and* (4.11) *it can be expected that the Hermite polynomials in* (4.11) *have a statistical interpretation as permanents of sampling matrices.*

Proposition 4.3. *Under the condition $\mu = E\xi_j = 0$ the normalized permanent of a sampling matrix with Bernoulli random elements has the following representation*

$$\sigma^{-m}n^{-m/2}\mathrm{Per}A_n^m(\xi) = H_m(\tau_{n1}, \tau_{n2}) + \Delta_n, \tag{4.12}$$

where $\Delta_n \xrightarrow{P} 0$ as $n \to \infty$.

Proof. The normalized permanent can be represented by the Cauchy integral as the term of the generating sampling function

$$\sigma^{-m}n^{-m/2}\mathrm{Per}A_n^m(\xi) = \frac{m!}{2\pi i}\oint \frac{dz}{z^{m+1}} \prod_{j=1}^{n}(1 + z\xi_j/\sigma\sqrt{n}). \tag{4.13}$$

The integral in (4.13) is taken along the circle with a small radius with the centre at the point $z = 0$. Unfortunately, the asymptotic representation (4.10) cannot be used because the residual factor depends on the integrable variable z. The more refined representation of the generating sampling function is obtained by means of expansion of the logarithm function in the Taylor series:

$$\prod_{j=1}^{n}(1 + z\xi_j/\sigma\sqrt{n}) = \exp[\sum_{j=1}^{n}\ln(1 + z\xi_j/\sigma\sqrt{n})]$$

$$= \exp[z\tau_{n1} - z^2\tau_{n2}/2]\exp[-\sum_{k=3}^{m}(-1)^k z^k \tau_{nk}/k]\exp[\varphi_n(z)], \qquad (4.14)$$

where

$$\varphi_n(z) := \sum_{k=m+1}^{\infty}(-1)^{k-1}z^k\tau_{nk}/k$$

and

$$\tau_{nk} := \sum_{j=1}^{n}\xi_j^k/\sigma^k n^{k/2}, \ k \geq 1.$$

By virtue of $|\tau_{nk}| \leq cn^{-1/2}$ for $k \geq 3$, the following representation is obtained

$$\sigma^{-m}n^{-m/2}\mathrm{Per}A_n^m(\xi)$$

$$= \frac{m!}{2\pi i}\oint\frac{dz}{z^{m+1}}\exp[z\tau_{n1} - z^2\tau_{n2}/2]\exp[-\sum_{k=3}^{m}(-1)^k z^k\tau_{nk}/k]$$

$$= \frac{m!}{2\pi i}\oint\frac{dz}{z^{m+1}}\exp[z\tau_{n1} - z^2\tau_{n2}/2] + \Delta_n.$$

It means that the representation (4.12) takes place.

Remark 4.3. *Under the condition $\mu = E\xi_j = 0$, by the law of large numbers,*

$$\tau_{n2} = \sum_{j=1}^{n}\xi_j^2/\sigma^2 n \xrightarrow{P} 1$$

as $n \to \infty$. So the permanent of a sampling matrix with Bernoulli distributed random elements can be interpreted as a statistical image of the Hermite polynomial as follows

$$\mathrm{Per}A_n^m(\xi/\sigma\sqrt{n}) = H_m(\tau_{n1}) + \Delta_n, \qquad (4.15)$$

where the Hermite polynomials $H_m(\tau)$ have the generating function

$$\exp[z\tau - z^2/2] = \sum_{m=0}^{\infty}\frac{z^m}{m!}H_m(\tau). \qquad (4.16)$$

In particular, the stochastic Kravchuk polynomial (4.5) can be interpreted as the statistical image of the Hermite polynomial as follows:

$$K_n^m(\nu_n, a) = H_m(\tau_{n1}) + \Delta_n, \tag{4.17}$$

where by the definition

$$K_n^m(\nu_n, a) = \sum_{r=0}^{m} a^r(-b)^{m-r}\nu_n^{[m]}(n-\nu_n)^{[m-r]}m^{[r]}/r! \ ,$$

$$\begin{aligned} \nu_n &:= \nu_{na} \\ &= \sum_{j=1}^{n} I(\xi_j = a). \end{aligned}$$

Here $\mu = E\xi_j = ap - bq = 0$.

4.2 Generating sampling function

Proposition 4.2 can be generalized for an arbitrary simple sample $\xi_j, 1 \le j \le n$, of independent and identically distributed random variables taking values in a measurable space $(R, \mathcal{X})$ with

$$\begin{aligned} E\xi_j &= \mu, \\ E(\xi_j - \mu)^2 &= \sigma^2 > 0, \\ v &= \mu/\sigma. \end{aligned} \tag{4.18}$$

Theorem 4.1. *Under the condition* (4.18) *the weak convergence, as* $n \to \infty$,

$$\prod_{j=1}^{n}(1 + z\xi_j/\sigma\sqrt{n})\exp[-zv\sqrt{n}] \Rightarrow \exp[z\tau - z^2(1+v^2)/2] \tag{4.19}$$

takes place for all $|z| \le z_0$.

Proof. At first note that, according to the proof of proposition 4.2, the statement (4.19) holds for all bounded random variables. In order to get (4.19) in the general case introduce truncated variables ($c > 0$):

$$\eta_j^c := \eta_j I(|\eta_j| \le c),$$

$$\bar{\eta}_j^c := \eta_j I(|\eta_j| > c) \ , \tag{4.20}$$

where $\eta_j := (\xi_j - \mu)/\sigma\sqrt{n} \ , 1 \le j \le n$.

The generating function with normalization factor can be now represented in the form

$$\Pi_n(z) = \prod_{j=1}^{n}(1 + z\xi_j/\sigma\sqrt{n})\exp[-zv\sqrt{n} + z^2v^2/2]$$

$$= \prod_{j=1}^{n}(1 + z\eta_j^c) + \Delta_{n1}(z) + \Delta_{n2}(z) .$$

Here,

$$\Delta_{n1}(z) = \prod_{j=1}^{n}(1 + z\eta_j) - \prod_{j=1}^{n}(1 + z\eta_j^c),$$

$$\Delta_{n2}(z) = \Pi_n(z) - \prod_{j=1}^{n}(1 + z\eta_j).$$

The limiting result in theorem 4.1 is a consequence of proposition 4.2 for bounded random variables and the following lemma.

Lemma 4.1. *Under the condition of theorem* 4.1, *the following estimates are valid*

$$E\Delta_{nk}^2(z) \longrightarrow 0 \quad \text{as} \quad n \to \infty, \quad k = 1, 2.$$

Proof. The following evident correlations are necessary

$$E(1 + z\xi_j/\sigma\sqrt{n})^2 = 1 + 2zv/\sqrt{n} + z^2(1 + v^2)/n,$$

$$E(1 + z\xi_j/\sigma\sqrt{n})(1 + z\eta_j) = 1 + zv/\sqrt{n} + z^2/n,$$

$$E(1 + z\eta_j)^2 = 1 + z^2/n,$$

$$E(1 + z\eta_j)(1 + z\eta_j^c) = 1 + (z^2 + \mathcal{E}_n(z))/n,$$

$$E(1 + z\eta_j^c/\sigma\sqrt{n})^2 = 1 + (z^2 + \mathcal{E}_n(z))/n.$$

Here $\mathcal{E}_n(z) \to 0$ as $n \to \infty$; in general, it is distinct in different formulas. Now compute the asymptotics of these moments

$$E\prod_{j=1}^{n}(1 + z\eta_j)^2 = (1 + z^2/n)^n$$

$$= e^{z^2}(1 + \mathcal{E}_n(z))$$

and by analogy,

$$E\prod_{j=1}^{n}(1+z\eta_j^c)^2 = e^{z^2}(1+\mathcal{E}_n(z)),$$

$$E\prod_{j=1}^{n}(1+z\eta_j)(1+z\eta_j^c) = e^{z^2}(1+\mathcal{E}_n(z)),$$

$$E\Pi_n^2(z) = e^{z^2}(1+\mathcal{E}_n(z)),$$

$$E\Pi_n(z)\prod_{j=1}^{n}(1+z\eta_j) = e^{z^2}(1+\mathcal{E}_n(z)).$$

Here, as above, $\mathcal{E}_n(z) \to 0$ as $n \to \infty$. These asymptotic results permit the conclusion that the statement of lemma 4.1 holds.

4.3 Generating multiple function

Consider a multiple sample

$$\xi_{sj}, \quad 1 \le s \le k, \quad 1 \le j \le n, \tag{4.21}$$

of random variables ξ_{sj} given on a measurable space $(R, \mathcal{X})$ which are jointly independent and identically distributed, for fixed s , and have finite mean values and variations, namely,

$$\begin{aligned} E\xi_{sj} &= \mu_s\,, \\ E(\xi_{sj}-\mu_s)^2 &= \sigma_s^2 > 0\,, \\ v_s &:= \mu_s/\sigma_s\,. \end{aligned} \tag{4.22}$$

The *generating multiple sampling function* is determined as

$$\Pi_n^{\xi}(z) := \prod_{j=1}^{n}\left[1+\sum_{s=1}^{k} z_s\xi_{sj}\right]. \tag{4.23}$$

Let τ_s, $1 \le s \le k$, be mutually independent standard normally distributed random variables.

Theorem 4.2. *Under the condition* (4.22) *of the finiteness of variations of sampling elements* (4.21), *the weak convergence*

$$\prod_{j=1}^{n}\left[1+\sum_{s=1}^{k} z_s\xi_{sj}/\sigma_s\sqrt{n}\right]\exp\left[-\sqrt{n}\sum_{s=1}^{k} z_s v_s\right]$$

$$\Rightarrow \exp\left[\sum_{s=1}^{k}(z_s\tau_s - z_s^2/2)\right]$$

$$\times \exp[-\left(\sum_{s=1}^{k} z_s v_s\right)^2/2] \tag{4.24}$$

takes place for all $|z| \leq z_0$ as $n \to \infty$, *where* z_0 *is a fixed positive number.*

Remark 4.4. *The exponentional function in the right hand-side of* (4.24) *is the generating function of the Hermite polynomials of k variables:*

$$\exp\left[\sum_{s=1}^{k}(z_s\tau_s - z_s^2/2)\right]$$

$$= \sum_{m=0}^{\infty} \sum_{m_1+\ldots+m_k=m} \prod_{s=1}^{k} \frac{z_s^{m_s}}{m_s!} H_{m_s}(\tau_s). \tag{4.25}$$

Proof. The proof is analogous to that of theorem 4.1 but involves a somewhat more complicated calculation For simplicity, suppose that z is a real number. First introduce the truncated variables

$$\xi_{sj}^c := \xi_{sj} I(|\xi_{sj}| \leq c\sqrt{n}),$$

$$\bar{\xi}_{sj}^c := \xi_{sj} I(|\xi_{sj}| > c\sqrt{n}) \ .$$

Consider the logarithm of the generating function of truncated variables with the normalization factor

$$\Pi_n^c := \ln \prod_{j=1}^{n}\left[1 + \sum_{s=1}^{k} z_s \xi_{sj}^c/\sigma_s\sqrt{n}\right] - \sqrt{n}\sum_{s=1}^{k} z_s v_s$$

$$= \sum_{j=1}^{n} \ln\left[1 + \sum_{s=1}^{k} z_s \xi_{sj}^c/\sigma_s\sqrt{n}\right] - \sqrt{n}\sum_{s=1}^{k} z_s v_s$$

$$= \sum_{s=1}^{k} z_s \tau_{sn}^c/\sigma_s\sqrt{n} - \sum_{j=1}^{n}\left[\sum_{s=1}^{k} z_s \xi_{sj}^c/\sigma_s\sqrt{n}\right]^2/2 - \sqrt{n}\sum_{s=1}^{k} z_s v_s + \Delta_n, \tag{4.26}$$

where

$$\tau_{sn}^c := \sum_{j=1}^{n} \xi_{sj}^c.$$

By using the equality $\xi_{sj} = \xi_{sj}^c + \bar{\xi}_{sj}^c$, it can be obtained that

$$\sum_{s=1}^{k} z_s \tau_{sn}^c/\sigma_s\sqrt{n} - \sqrt{n}\sum_{s=1}^{k} z_s v_s$$

$$= \sum_{s=1}^{k} z_s \tau_{sn}^0 / \sigma_s \sqrt{n} - \sum_{s=1}^{k} z_s \bar{\tau}_{sn}^c / \sigma_s \sqrt{n},$$

where

$$\tau_{sn}^0 := \sum_{j=1}^{n} (\xi_{sj} - \mu_s),$$

$$\bar{\tau}_{sn}^c := \sum_{j=1}^{n} \bar{\xi}_{sj}^c.$$

Just as in the proof of theorem 4.1 , by the central limit theorem, obtain

$$\sum_{s=1}^{k} z_s \tau_{sn}^0 / \sigma_s \sqrt{n} \Longrightarrow \sum_{s=1}^{k} z_s \tau_s \quad \text{as} \quad n \to \infty,$$

and

$$\sum_{s=1}^{k} z_s \bar{\tau}_{sn}^c / \sigma_s \sqrt{n} \xrightarrow{P} 0 \quad \text{as} \quad n \to \infty,$$

because

$$E[\bar{\tau}_{sn}^c / \sqrt{n}]^2 \longrightarrow 0 \quad \text{as} \quad n \to \infty.$$

For the second term in (4.26) obtain

$$\sum_{j=1}^{n} \left(\sum_{s=1}^{k} z_s \xi_{sj}^c / \sigma_s \sqrt{n} \right)^2$$

$$= \sum_{s=1}^{k} z_s^2 (\tau_{sn}^c)^2 / \sigma_s^2 n + 2 \sum_{1 \le s_1 < s_2 \le k} z_{s_1} z_{s_2} \sum_{j=1}^{n} \xi_{s_1 j}^c \xi_{s_2 j}^c / \sigma_{s_1} \sigma_{s_2} n.$$

By the law of large numbers,

$$\sum_{s=1}^{k} z_s^2 (\tau_{sn}^c)^2 / \sigma_s^2 n \xrightarrow{P} \sum_{s=1}^{k} z_s^2 (1 + v_s^2),$$

and

$$\sum_{1 \le s_1 < s_2 \le k} z_{s_1} z_{s_2} \sum_{j=1}^{n} \xi_{s_1 j}^c \xi_{s_2 j}^c / \sigma_{s_1} \sigma_{s_2} n \xrightarrow{P} \sum_{1 \le s_1 < s_2 \le k} z_{s_1} z_{s_2} v_{s_1} v_{s_2}$$

as $n \to \infty$.

Consequently,

$$\sum_{j=1}^{n} \left(\sum_{s=1}^{k} z_s \xi_{sj}^c / \sigma_s \sqrt{n} \right)^2 \xrightarrow{P} \sum_{s=1}^{k} z_s^2 + \left(\sum_{s=1}^{k} z_s v_s \right)^2 \quad \text{as} \quad n \to \infty. \tag{4.27}$$

By analogy with the proof of theorem 4.1, employing the estimate (4.23) of the logarithm function and the Marzinkevich law of large numbers (Loeve (1960)), obtain the following estimate of the last term in (4.26):

$$\Delta_n \xrightarrow{P} 0 \quad \text{as} \quad n \to \infty . \tag{4.28}$$

Gathering the estimates obtained above and taking into account (4.26) - (4.28),

$$\begin{aligned} \exp(\Pi_n^c) \;\Rightarrow\; & \exp\left[\sum_{s=1}^{k}(z_s\tau_s - z_s^2/2)\right] \\ & \times \exp[-\left(\sum_{s=1}^{k} z_s v_s\right)^2/2] \quad \text{as} \quad n \to \infty . \end{aligned}$$

Consider the value

$$\Delta_n := \left(\prod_{j=1}^{n}\left[1+\sum_{s=1}^{k} z_s\xi_{sj}/\sigma_s\sqrt{n}\right] - \prod_{j=1}^{n}\left[1+\sum_{s=1}^{k} z_s\xi_{sj}^c/\sigma_s\sqrt{n}\right]\right)\exp(-\sqrt{n}\sum_{s=1}^{k} z_s v_s).$$

First compute

$$\begin{aligned} a_{n2} &:= E\left[1+\sum_{s=1}^{k} z_s\xi_{s1}^c/\sigma_s\sqrt{n}\right]^2 \\ &= a_{n1} + \mathcal{E}_{n1}(z)/n, \end{aligned}$$

where

$$\begin{aligned} a_{n1} &:= E\left[1+\sum_{s=1}^{k} z_s\xi_{s1}/\sigma_s\sqrt{n}\right]^2 \\ &= 1+2\sum_{s=1}^{k} z_s v_s/\sqrt{n} + \sum_{s=1}^{k} z_s^2/n \\ &\quad +\left(\sum_{s=1}^{k} z_s v_s\right)^2/n, \end{aligned}$$

$$\begin{aligned} \mathcal{E}_{n1}(z) &:= E\left[\sum_{s=1}^{k} z_s\cdot\bar{\xi}_{s1}^c/\sigma_s\right]^2 - 2\sqrt{n}\sum_{s=1}^{k} z_s(E\bar{\xi}_{s1}^c)/\sigma_s \\ &\quad -2E(\sum_{s=1}^{k} z_s\xi_{s1}/\sigma_s)(\sum_{s=1}^{k} z_s\bar{\xi}_{s1}^c/\sigma_s) \end{aligned}$$

Compute also

$$\begin{aligned} a_{n3} &:= E\left(1+\sum_{s=1}^{k} z_s\xi_{s1}/\sigma_s\sqrt{n}\right)\left(1+\sum_{s=1}^{k} z_s\xi_{s1}^{c}/\sigma_s\sqrt{n}\right) \\ &= a_{n1}+\mathcal{E}_{n2}(z)/n, \end{aligned}$$

where

$$\mathcal{E}_{n2}(z) := -\sqrt{n}\sum_{s=1}^{k} z_s(E\bar{\xi}_{s1}^{c})/\sigma_s - E(\sum_{s=1}^{k} z_s\xi_{s1}/\sigma_s)(\sum_{s=1}^{k} z_s\bar{\xi}_{s1}^{c}/\sigma_s)$$

Under the conditions of the theorem

$$\mathcal{E}_{nr}(z)\to 0, \quad r=1,2, \quad \text{as} \quad n\to\infty.$$

Now complete the proof of theorem 4.2, as in that of theorem 4.1. By analogy with lemma 4.1, obtain the asymptotic estimates

$$n\ln a_{nr} - 2\sum_{s=1}^{k} z_s v_s = \sum_{s=1}^{k} z_s^2 - (\sum_{s=1}^{k} z_s v_s)^2 + \Delta_{nr}, \quad r=1,2,3,$$

where $\Delta_{nr} \xrightarrow{P} 0$ as $n\to\infty$, $r=1,2,3$.

Finally,

$$\begin{aligned} E|\Delta_n|^2 := E&\left|\prod_{j=1}^{n}\left[1+\sum_{s=1}^{k} z_s\xi_{sj}/\sigma_s\sqrt{n}\right] - \prod_{j=1}^{n}\left[1+\sum_{s=1}^{k} z_s\xi_{sj}^{c}/\sigma_s\sqrt{n}\right]\right|^2 \\ &\times\exp\left[-2\sqrt{n}\sum_{s=1}^{k} z_s v_s\right] \\ &= (a_{n1}^{n}+a_{n2}^{n}-2a_{n3}^{n})\exp\left[-2\sqrt{n}\sum_{s=1}^{k} z_s v_s\right] \\ &= (e^{\Delta_{n1}}+e^{\Delta_{n2}}-2e^{\Delta_{n3}}) \\ \times\exp&\left(\sum_{s=1}^{k} z_s^2 - \left(\sum_{s=1}^{k} z_s v_s\right)^2\right) \to 0, \quad \text{as} \quad n\to\infty. \end{aligned}$$

Hence $\Delta_n \xrightarrow{P} 0$ as $n\to\infty$. Thus the proof of theorem 4.2 is completed.

Let $\tau_s, 1\le s\le k$ be a family of normally distributed random variables with the generating function

$$E\exp\left[\sum_{s=1}^{k} z_s\tau_s\right] = \exp\left[\sum_{s,r=1}^{k} \rho_{sr}z_s z_r/2\right],$$

where

$$\rho_{sr} := E((\tau_s - \mu_s)(\tau_r - \mu_r))/\sigma_s\sigma_r,$$

$$\sigma_s^2 := E(\tau_s - \mu_s)^2,$$
$$\mu_s := E\tau_s \ ,$$
$$v_s := \mu_s/\sigma_s \ .$$

Consider the multiple sample $\xi_{sj}, 1 \leq s \leq k \ , \ 1 \leq j \leq n$ of independent and identically distributed random variables for a fixed s, having the same first two moments as the vector $\tau = (\tau_s, \ 1 \leq s \leq k \ )$:

$$E(\xi_{sj} - \mu_s)(\xi_{rj} - \mu_r)/\sigma_s\sigma_r = \rho_{sr},$$

$$E(\xi_{sj} - \mu_s)^2 = \sigma_s^2,$$

$$E\xi_{sj} = \mu_s.$$

Theorem 4.3. *The normalized generating multiple sampling function is weak convergent as $n \to \infty$:*

$$\prod_{j=1}^{n}\left(1 + \sum_{s=1}^{k} z_s\xi_{sj}/\sigma_s\sqrt{n}\right)\exp\left[-\sqrt{n}\sum_{s=1}^{k} z_s v_s\right]$$
$$\Rightarrow \exp\left[\sum_{s=1}^{k} z_s\tau_s\right]/E\exp\left[\sum_{s=1}^{k} z_s\tau_s\right] \tag{4.29}$$

to the generating function of the Appell–Wick polynomials :

$$\exp\left[\sum_{s=1}^{k} z_s\tau_s\right]/E\exp\left[\sum_{s=1}^{k} z_s\tau_s\right]$$
$$= \sum_{m=0}^{\infty}\sum_{m_1+\ldots+m_k=m}\prod_{s=1}^{k}\frac{z_s^{m_s}}{m_s!} : \tau_1^{m_1}\ldots\tau_k^{m_k} : . \tag{4.30}$$

In particular, for $\rho_{sr} = 0$ with $s \neq r$, the Appell–Wick polynomials have the following representation:

$$: \tau_1^{m_1}\ldots\tau_k^{m_k} : = \prod_{s=1}^{k} H_{m_s}(\tau_s) \tag{4.31}$$

Proof. Introduce the truncated random variables

$$\xi_{sj}^c = \xi_{sj}I(|\xi_{sj}| \leq c\sigma_s\sqrt{n}), 0 < c < 1.$$

Under the conditions of the theorem

$$\left(\prod_{j=1}^{n}(1+\sum_{s=1}^{k} z_s\xi_{sj}/\sigma_s\sqrt{n}) - \prod_{j=1}^{n}(1+\sum_{s=1}^{k} z_s\xi_{sj}^c/\sigma_s\sqrt{n})\right)\exp(-\sqrt{n}\sum_{s=1}^{k} z_s v_s) \xrightarrow{P} 0.$$

This statement is proved with the help of arguments which are outlined in the deduction of analogous estimates in the proof of theorem 4.2.

Further, with the help of the Taylor formula for logarithmic functions, the generating multiple sampling function with truncated variables is represented in the form

$$\begin{aligned}\prod_{j=1}^{n}(1 &+ \sum_{s=1}^{k} z_s\xi_{sj}^c/\sigma_s\sqrt{n}) \\ &= \exp\left(\sum_{j=1}^{n}\ln(1+\sum_{s=1}^{k} z_s\xi_{sj}^c/\sigma_s\sqrt{n})\right) \\ &= \exp\left(\sum_{r=1}^{\infty}(-1)^{r+1}\tau_{nr}^c/r\right),\end{aligned}$$

where

$$\tau_{nr}^c = \sum_{j=1}^{n}\left(\sum_{s=1}^{k} z_s\xi_{sj}^c/\sigma_s\sqrt{n}\right)^r, \quad r \geq 1.$$

Assume that

$$kz_0c = \delta < 1.$$

For $r \geq 3$

$$\begin{aligned}|\tau_{nr}^c| &\leq \delta^{r-3}\sum_{j=1}^{n}\left|\sum_{s=1}^{k} z_s\xi_{sj}^c/\sigma_s\sqrt{n}\right|^3 \\ &\leq k^2\delta^{r-3}\sum_{s=1}^{k}|z_s|^3\sigma_s^{-3}\left(n^{-3/2}\sum_{j=1}^{n}|\xi_{sj}^c|^3\right).\end{aligned}$$

According to the Marzinkevich law of large numbers with probability 1

$$n^{-3/2}\sum_{j=1}^{n}|\xi_{sj}^c|^3 \to 0, \quad n\to\infty, \quad 1\leq s\leq k,$$

when $\sigma_s^2 < \infty$, $s = 1,\dots,k$.

Consequently

$$\left|\sum_{r=3}^{\infty}(-1)^{r+1}\tau_{nr}^c/r\right|$$

$$\leq k^2 \cdot \sum_{s=1}^{k} |z_s|^3 \sigma_s^{-3} \left(n^{-3/2} \sum_{j=1}^{n} |\xi_{sj}^c|^3 \right) \left(\sum_{r=3}^{\infty} \delta^{r-3}/r \right)$$

$$\leq \ln(1-\delta)^{-1} k^2 \cdot \sum_{s=1}^{k} |z_s|^3 \sigma_s^{-3} \left(n^{-3/2} \sum_{j=1}^{n} |\xi_{sj}^c|^3 \right) \to 0$$

with probability 1, when $n \to \infty$.

According to the strong law of large numbers

$$n^{-1} \sum_{j=1}^{n} \xi_{s_1 j}^c \xi_{s_2 j}^c \to E(\xi_{s_1 j} \xi_{s_2 j})$$

almost surely.

Therefore, when $n \to \infty$,

$$\tau_{n2}^c \to \sum_{s_1, s_2 = 1}^{k} \rho(s_1, s_2) z_{s_1} z_{s_2}$$

almost surely.

Further, according to the multidimensional central limit theorem, the weak convergence

$$\tau_{n1}^c - \sqrt{n} \sum_{s=1}^{k} z_s v_s$$

$$= \sum_{s=1}^{k} z_s \sum_{j=1}^{n} (\xi_{sj}^c - \mu_s)/\sigma_s \sqrt{n} \Rightarrow \sum_{s=1} z_s \tau_s$$

takes place.

As a result the weak convergence

$$\prod_{j=1}^{n} (1 + \sum_{s=1}^{k} z_s \xi_{sj}^c / \sigma_s \sqrt{n}) \exp(-\sqrt{n} \sum_{s=1}^{k} z_s v_s)$$

$$\Rightarrow \exp(\sum_{s=1}^{k} z_s \tau_s) / E \exp(\sum_{s=1}^{k} z_s \tau_s)$$

takes place and is in accordance with (4.29).

4.4 Sampling permanents

Proposition 4.3 for the permanent of a sampling matrix with Bernoulli distributed random elements can be generalized on an arbitrary simple sample with centered variables.

Theorem 4.4. *Under the condition*

$$E\xi_j = 0,$$

$$E\xi_j^2 = \sigma^2 > 0 \ , \ 1 \le j \le n \, ,$$

the weak convergence

$$\sigma^{-m} n^{-m/2} \mathrm{Per} A_n^m(\xi) \Rightarrow H_m(\tau) \quad \text{as} \quad n \to \infty,$$

takes place.

Proof. By virtue of the Cauchy formula for residues, the following representation of permanents in terms of the generating function is available

$$\sigma^{-m} n^{-m/2} \mathrm{Per} A_n^m(\xi) = \frac{m!}{2\pi i} \oint \frac{dz}{z^{m+1}} \prod_{j=1}^{n} (1 + z\xi_j / \sigma\sqrt{n}). \tag{4.32}$$

The integral in (4.32) is taken along the circle with a very small radius circumscribed around the point $z = 0$. In order to use the logarithm of the generating function, apply the truncated variables

$$\xi_j^c := \xi_j I(|\xi_j| \le c\sqrt{n}),$$

$$\bar{\xi}_j^c := \xi_j I(|\xi_j| > c\sqrt{n}).$$

As in the proof of theorem 4.1 write

$$\sigma^{-m} n^{-m/2} \mathrm{Per} A_n^m(\xi) = \frac{m!}{2\pi i} \oint \frac{dz}{z^{m+1}} \prod_{j=1}^{n} (1 + z\xi_j^c / \sigma\sqrt{n}) + \Delta_n,$$

where

$$\Delta_n := \frac{m!}{2\pi i} \oint \frac{dz}{z^{m+1}} \Delta_n(z),$$

$$\Delta_n(z) := \prod_{j=1}^{n} (1 + z\xi_j / \sigma\sqrt{n}) - \prod_{j=1}^{n} (1 + z\xi_j^c / \sigma\sqrt{n}) \, .$$

In the proof of theorem 4.1 it has been established that

$$\Delta_n(z) \xrightarrow{P} 0 \ \text{ as } n \to \infty \, .$$

Hence $\Delta_n \xrightarrow{P} 0$ as $n \to \infty$. By using the limiting result of theorem 4.1, it is possible to take the weak limit in

$$\begin{aligned} \sigma^{-m} n^{-m/2} \mathrm{Per} A_n^m(\xi) \quad &\Rightarrow \quad \frac{m!}{2\pi i} \oint \frac{dz}{z^{m+1}} \exp[z\tau - z^2/2] \\ &= \quad H_m(\tau) \, . \end{aligned}$$

Remark 4.5. *The definition of a permanent as the term of the generating sampling function by the Cauchy integral* (4.32) *allows a precise representation in the following form*

$$\sigma^{-m}n^{-m/2}\mathrm{Per}A_n^m(\xi) = H_m(\tau_{n1},\tau_{n2}) + \sum_{k=3}^{m} P_k(\tau_n^{(k)})H_{m-k}(\tau_{n1},\tau_{n2})m^{[k]}, \qquad (4.33)$$

where $P_k(\tau_n^{(k)})$ *is the polynomial of degree* k *with variables* $\tau_n^{(k)} := (\tau_{n3},\ldots,\tau_{nk})$, *which are defined in the asymptotic expansion*

$$\exp[-\sum_{k=3}^{m}(-1)^k z^k \tau_{nk}/k] = 1 + \sum_{k=3}^{\infty} z^k P_k(\tau_n^{(k)}).$$

Previously

$$\tau_{nk} := \sum_{j=1}^{n} \xi_j^k/\sigma^k n^{k/2}.$$

So, by the strong law of large numbers,

$$\tau_{nk} \to 0 \;\text{ as }\; n \to \infty \;\text{ for }\; k \geq 3$$

with probability 1.

For the proof of (4.33) use, for the beginning, the formula

$$\mathrm{Per}A_n^m(\xi) = \lim_{c\to\infty} \mathrm{Per}A_n^m(\xi^c),$$

where

$$A_n^m(\xi^c) = [a_{ij} = \xi_j^c;\; 1 \leq i \leq m,\; 1 \leq j \leq n],$$

and then apply for $\mathrm{Per}A_n^m(\xi^c)$ with arbitrary fixed $c > 0$ the reasoning from proposition 4.3 for bounded random variables, choosing in (4.13) a contour of integration inside the domain $|z| < \sigma c^{-1}$. In this domain along the contour of integration the validity of the decomposition (4.14) is provided.

4.5 Multiple sampling permanents

As in section 4.3 consider the multiple sample

$$\xi_{sj}, \quad 1 \leq s \leq k, \quad 1 \leq j \leq n$$

of random variables ξ_{sj} given on a measurable space $(R, \mathcal{X})$ which are jointly independent and identically distributed for fixed s, and have zero mean values and finite variations, namely,

$$E\xi_{sj} = 0,$$

$$E\xi_{sj}^2 = \sigma_s^2 > 0 \quad, 1 \le s \le k, \quad 1 \le j \le n \,. \tag{4.34}$$

Introduce a multiple sampling matrix

$$A_n^m(\xi_1, \ldots, \xi_k) := \begin{bmatrix} A_n^{m_1}(\xi_1) \\ \cdots \\ A_n^{m_k}(\xi_k) \end{bmatrix},$$

where

$$A_n^{m_s}(\xi_s) := [a_{ij} = \xi_{sj} \;; 1 \le i \le m_s, \quad 1 \le j \le n], \quad s = 1, \ldots, k,$$

are sampling matrices. Here, $m_1 + \ldots + m_k = m$.

Let τ_s, $1 \le s \le k$, be standard mutually independent and normally distributed random variables. The method of proving theorem 4.4 can be applied in a more general case to the permanent of multiple sampling matrices.

Theorem 4.5. *Under the condition* (4.34), *the weak convergence*

$$n^{-m/2} \prod_{s=1}^{k} \sigma_s^{-m_s} \mathrm{Per} A_n^m(\xi_1, \ldots, \xi_k) \Rightarrow \prod_{s=1}^{k} H_{m_s}(\tau_s) \tag{4.35}$$

takes place as $n \to \infty$.

Proof. Taking into account the result of theorem 1.2, by virtue of the Cauchy formula for residues,

$$J_n := n^{-m/2} \prod_{s=1}^{k} \sigma_s^{-m_s} \mathrm{Per} A_n^m(\xi_1, \ldots, \xi_k)$$

$$= \frac{1}{(2\pi i)^k} \prod_{s=1}^{k} m_s! \oint \cdots \oint \prod_{s=1}^{k} \frac{dz_s}{z^{m_s+1}} \prod_{j=1}^{n} \left[1 + \sum_{s=1}^{k} z_s \xi_{sj} / \sigma_s \sqrt{n} \right] \,. \tag{4.36}$$

The integrals in (4.36) are taken along the circles with small radius circumscribed around the points $z_s = 0$, $1 \le s \le k$.

Comparing formula (4.36) with (4.32) and taking into account the limiting result (4.17) in theorem 4.2, the result (4.35) can be established.

In order to obtain the exact result, consider the truncated random variables

$$\xi_{sj}^c := \xi_{sj} I(|\xi_{sj}| \le c\sqrt{n}),$$

$$\bar{\xi}_{sj}^c := \xi_{sj} I(|\xi_{sj}^c| > c\sqrt{n}).$$

As in the proof of theorem 4.4 for the permanents of the truncated variables,

$$J_n^c = n^{-m/2} \prod_{s=1}^{k} \sigma_s^{-m_s} \mathrm{Per} A_n^m(\xi_1^c, \ldots, \xi_k^c)$$

$$= (2\pi i)^{-k} \prod_{s=1}^{k} m_s! \oint \ldots \oint \prod_{s=1}^{k} \frac{dz_s}{z_s^{m_s+1}} \exp \sum_{j=1}^{n} \ln \left[1 + \sum_{s=1}^{k} z_s \xi_{sj}^c / \sigma_s \sqrt{n} \right]$$

$$= (2\pi i)^{-k} \prod_{s=1}^{k} m_s! \oint \ldots \oint \prod_{s=1}^{k} \frac{dz_s}{z_s^{m_s+1}} \exp \left[\sum_{s=1}^{k} (z_s \tau_{ns}^c / \sigma_s \sqrt{n} - z_s^2 \tau_{ns}^c / \sigma_s^2 n) \right]$$

$$\times \exp \left[\sum_{r=3}^{m} (-1)^r \sum_{s=1}^{k} z_s^r \tau_{nsr}^c / r \sigma_s^{r/2} n^{r/2} \right],$$

where

$$z_s^r := \prod_{e=1}^{r} z_{s_e}, \quad \tau_{nsr}^c := \sum_{j=1}^{n} \prod_{e=1}^{r} \xi_{s_e j}, \quad r \geq 3 .$$

By the central limit theorem,

$$\tau_{ns}^c / \sigma_s \sqrt{n} \Rightarrow \tau_s , \quad n \to \infty.$$

By the law of large numbers,

$$\tau_{ns}^c / 2\sigma_s^2 n \to 1/2 , \quad \text{(a.s.)} \quad n \to \infty.$$

And by the Marzinkevich law of large numbers,

$$\tau_{nsr}^c / n^{r/2} \to 0, \quad \text{(a.s.)} \quad n \to \infty, \quad r \geq 3.$$

Hence, using the generating formula for Hermite polynomials (4.18),

$$J_n^c \Rightarrow \frac{1}{(2\pi i)^k} \prod_{s=1}^{k} m_s! \oint \ldots \oint \prod_{s=1}^{k} \frac{dz_s}{z_s^{m_s+1}} \exp \left[\sum_{s=1}^{k} (z_s \tau_s - z_s^2/2) \right]$$

$$= \prod_{s=1}^{k} H_{m_s}(\tau_s)$$

as $n \to \infty$. To complete the proof of theorem 4.6 consider the difference

$$\begin{aligned} \Delta_n &:= J_n - J_n^c \\ &= (2\pi i)^{-k} \prod_{s=1}^{k} m_s! \oint \ldots \oint \prod_{s=1}^{k} \frac{dz_s}{z^{m_s+1}} \Delta_n(z), \end{aligned}$$

where

$$\Delta_n(z) := \prod_{j=1}^{n} (1 + \sum_{s=1}^{k} z_s \xi_{sj} / \sigma_s \sqrt{n}) - \prod_{j=1}^{n} (1 + \sum_{s=1}^{k} z_s \xi_{sj}^c / \sigma_s \sqrt{n}).$$

The same reasons as in the proof of theorem 4.4 enable the derivation

$$E|\Delta_n(z)|^2 \to 0 \quad \text{as} \quad n \to \infty.$$

This means that

$$\Delta_n(z) \xrightarrow{P} 0 \quad \text{as} \quad n \to \infty.$$

Remark 4.6. *By virtue of theorem 4.5 the permanent of a multiple sampling matrix with independent random elements in rows can be considered as a statistical image of the product of Hermite polynomials:*

$$n^{-m/2} \prod_{s=1}^{k} \sigma_s^{-m_s} \mathrm{Per} A_n^m(\xi_1, \ldots, \xi_k) = \prod_{s=1}^{k} H_{m_s}(\tau_n(s)) + \Delta_n,$$

where $\Delta_n \xrightarrow{P} 0$ *as* $n \to \infty$,

$$\tau_n(s) = n^{-1/2} \sigma_s^{-1} \sum_{j=1}^{n} \xi_{sj}.$$

In the more general case of a multiple sampling matrix with dependent elements in rows this can be interpreted as a statistical image of Appell–Wick polynomials.

Theorem 4.6. *For a multiple sample* ξ_{sj}, $1 \leq s \leq k$, $1 \leq j \leq n$, *with dependent samples in* s:

$$E\xi_{sj} = 0,$$

$$E\xi_{sj}^2 = \sigma_s^2 < \infty,$$

$$E(\xi_{sj}\xi_{rj})/\sigma_s\sigma_r = \rho_{sr}.$$

The permanent of a multiple sample matrix is weakly convergent, as $n \to \infty$, *to the Appell–Wick polynomial defined in theorem 4.3, namely:*

$$n^{-m/2} \prod_{s=1}^{k} \sigma_s^{-m_s} \mathrm{Per} A_n^m(\xi_1, \ldots, \xi_k) \ \Rightarrow \ : \tau_1^{m_1} \ldots \tau_k^{m_k} : .$$

Proof. First represent the permanent as a Cauchy integral

$$\mathrm{Per}\ A_n^m(\xi_1/\sigma_1\sqrt{n}, \ldots, \xi_k/\sigma_k\sqrt{n})$$

$$= \frac{m_1! \cdots m_k!}{(2\pi i)^k} \oint \cdots \oint \frac{dz_1}{z^{m_1+1}} \cdots \frac{dz_k}{z^{m_k+1}}$$

$$\times \prod_{j=1}^{n} (1 + \sum_{s=1}^{k} z_s \xi_{sj}/\sigma_s\sqrt{n}), \tag{4.38}$$

where the integration in each variable z_s, $s = 1, \ldots, k$, is taken along the circle of small radius which contains the point $z_s = 0$ and the conditions

$$m_1 + \cdots + m_k = m, \quad 0 \le m \le n,$$

$$0 \le m_s \le n, \quad s = 1, \ldots, k.$$

In order to eliminate the integrals in (4.38), transform the integrand function with the help of a power series to a suitable form.

With application of the Taylor series for a logarithmic function, for fixed sampling ξ_s, $s = 1, \ldots, k$,

$$\prod_{j=1}^{n}(1 + \sum_{s=1}^{k} z_s \xi_{sj}/\sigma_s\sqrt{n})$$

$$= \exp\left(\sum_{s_1=1}^{k} z_{s_1}\tau_n(s_1) - \sum_{s_1=1}^{k}\sum_{s_2=1}^{k} z_{s_1}z_{s_2}\tau_n(s_1, s_2)/2\right.$$

$$+ \sum_{r=3}^{m}(-1)^{r-1}r^{-1}\sum_{j=1}^{n}(\sum_{s=1}^{k} z_s\xi_{sj}/\sigma_s\sqrt{n})^r$$

$$\left. + \sum_{r=m+1}^{\infty}(-1)^{r-1}r^{-1}\sum_{s_1=1}^{k}\cdots\sum_{s_r=1}^{k} z_{s_1}\ldots z_{s_r}\tau_n(s_1, \ldots, s_r)\right), \tag{4.39}$$

where

$$\tau_n(s_1, \ldots, s_r) = n^{-r/2}\sum_{j=1}^{n} \xi_{s_1 j}\ldots\xi_{s_r j}/\sigma_{s_1}\cdots\sigma_{s_r}.$$

Realizing the substitution of (4.39) into (4.38) and taking into account the properties of Cauchy integrals,

$$\text{Per } A_n^m(\xi_1/\sigma_1\sqrt{n}, \ldots, \xi_k/\sigma_k\sqrt{n})$$

$$= \frac{m_1!\cdots m_k!}{(2\pi i)^k}\oint\cdots\oint\prod_{s=1}^{k}\frac{dz_s}{z^{m_s+1}}\exp(\sum_{s_1=1}^{k} z_{s_1}\tau_n(s_1)$$

$$-\sum_{s_1=1}^{k}\sum_{s_2=1}^{k} z_{s_1}z_{s_2}\tau_n(s_1, s_2)$$

$$+\sum_{r=3}^{m}(-1)^{r-1}r^{-1}\sum_{j=1}^{n}(\sum_{s=1}^{k} z_s\xi_{sj}/\sigma_s\sqrt{n})^r). \tag{4.40}$$

According to the central limit theorem

$$\tau_n(s_1) \Rightarrow \tau_{s_1},$$

and the strong law of large numbers

$$\tau_n(s_1, s_2) \to \rho_{s_1 s_2}$$

with probability 1.

Consider the sum in r in (4.40) separately. According to the formula

$$(\sum_{s=1}^{k} a_s)^r = \sum_{r_1+\cdots+r_k=r} \frac{r!}{r_1!\ldots r_k!}\prod_{s=1}^{k} a_s^{r_s}, \quad a_s \in R,$$

hence the representation

$$\sum_{j=1}^{n}(\sum_{s=1}^{k} z_s\xi_{sj}/\sigma_s\sqrt{n})^r$$

$$= \sum_{r_1+\cdots+r_k=r} \frac{r!}{r_1!\ldots r_k!}\prod_{s=1}^{k} z_s^{r_s}\sigma_s^{-r_s}\sum_{j=1}^{n}\prod_{s=1}^{k}\xi_{sj}^{r_s}n^{-r_s/2}$$

holds.

It is necessary to prove that, under the conditions of the theorem,

$$\sum_{j=1}^{k}\prod_{s=1}^{k}\xi_{sj}^{r_s}n^{-r_s/2} \to 0, \tag{4.41}$$

almost surely when $3 \le r \le m$. For this purpose apply the strong law of large numbers in Marzinkevich form.

Following the notations of Loeve (1960) put

$$p = 2/r,\ X_j = \xi_{1j}^{r_1}\xi_{2j}^{r_2}\ldots\xi_{kj}^{r_k}.$$

According to the Hölder inequality

$$\begin{aligned} E|\xi_{1j}^{r_1}\ldots\xi_{kj}^{r_k}|^{2/r} &= E(|\xi_{1j}^2|^{r_1/r}\ldots|\xi_{kj}^2|^{r_k/r}) \\ &\le (E\xi_{1j}^2)^{r_1/r}\ldots(E\xi_{kj}^2)^{r_k/r} \\ &= \prod_{s=1}^{k}\sigma_s^{2r_s/r} < \infty. \end{aligned}$$

That is why conditions of the theorem from (Loeve (1960)) are satisfied. This fact proves (4.41). Consequently, in (4.40) the relation (4.29) is true when $n \to \infty$. This means that the weak convergence

$$\operatorname{Per} A_n^m(\xi_1/\sigma_1\sqrt{n},\ldots,\xi_k/\sigma_1\sqrt{n})$$

$$\Rightarrow \frac{m_1!\ldots m_k!}{(2\pi i)^k}\oint\cdots\oint\prod_{s=1}^{k} dz_s z_s^{-m_s-1}$$

$$\times \exp(z_1\tau_1+\ldots+z_k\tau_k)/E\exp(z_1\tau_1+\cdots+z_k\tau_k)$$

occurs. From this fact and (4.30), the statement (4.37) follows.

4.6 Permanents of sampling matrices with increasing degrees

Consider a simple sample ξ_j , $1 \le j \le n$, of independent and identically distributed random variables which satisfy the conditions

$$\begin{aligned} E\xi_j &= \mu \neq 0 , \\ E(\xi_j - \mu) &= \sigma^2 > 0 , \\ v &= \mu/\sigma . \end{aligned} \tag{4.42}$$

Theorem 4.7. *Suppose that the simple sample satisfies the conditions* (4.42) *and the degree* m *of random permanents increases so that the following condition holds:*

$$b_n := m/\sqrt{n} \longrightarrow b > 0 \quad \text{as} \quad n \to \infty . \tag{4.43}$$

Then the weak convergence

$$n^{-[m]}\mu^{-m}\mathrm{Per}A_n^m(\xi) \Rightarrow \exp(a\tau - a^2/2) \tag{4.44}$$

takes place with

$$\begin{aligned} a &:= b/v \\ &= b\sigma/\mu . \end{aligned}$$

Proof. Setting $\eta := (\xi - \mu)/\sigma$, so that $\xi/\mu = 1 + \eta/v$, to get the formula

$$n^{-[m]}\mu^{-m}\mathrm{Per}A_n^m(\xi) = n^{-[m]}\mathrm{Per}A_n^m(1 + \eta/v).$$

By using the shift decomposition formula (1.22), obtain

$$n^{-[m]}\mathrm{Per}A_n^m(1 + \eta/v) = \sum_{c=0}^{m} \frac{m^{[c]}}{n^{[c]}v^c}\mathrm{Per}A_n^c(\eta)/c!$$

Now transform this sum as follows

$$\begin{aligned} &\sum_{c=0}^{m} \frac{m^{[c]}}{n^{[c]}v^c}\mathrm{Per}A_n^c(\eta)/c \\ &= \sum_{c=0}^{n} \left(\frac{m}{nv}\right)^c \mathrm{Per}A_n^c(\eta)/c! - \Delta_{n1} + \Delta_{n2} , \end{aligned} \tag{4.45}$$

where

$$\Delta_{n1} := \sum_{c=m+1}^{n} \left(\frac{m}{nv}\right)^c \mathrm{Per} A_n^c(\eta)/c!,$$

$$\Delta_{n2} := \sum_{c=0}^{m} \left(\frac{m}{nv}\right)^c \mathcal{E}_n(c) \mathrm{Per} A_n^c(\eta)/c!\,,$$

$$\mathcal{E}_n(c) := \prod_{s=1}^{c-1} \left(1 - \frac{s}{m}\right) \Big/ \left(1 - \frac{s}{n}\right) - 1.$$

By definition of the generating function of permanents (1.14) for the first term in (4.44)

$$\sum_{c=0}^{n} \left(\frac{m}{nv}\right)^c \mathrm{Per} A_n^c(\eta) = \prod_{j=1}^{n} (1 + b_n \eta_j / v\sqrt{n}).$$

By virtue of theorem 4.1, taking into account (4.43), obtain the following limiting result

$$\prod_{j=1}^{n} (1 + b_n \eta_j / v\sqrt{n}) \Rightarrow \exp(a\tau - a^2/2).$$

The validity of theorem 4.7 is now the consequence of the next lemma.

Lemma 4.2. *Under the conditions of theorem* 4.7 *the following asymptotic estimates hold*

$$E\Delta_{nk}^2 \longrightarrow 0 \quad \text{as} \quad n \to \infty, \quad k = 1,\ 2\ .$$

Proof. First note that for

$$q_s := \left(1 - \frac{s}{m}\right) \Big/ \left(1 - \frac{s}{n}\right)$$

the following inequality is true $1 - s/m \le q_s \le 1$.

By using (3.10) now compute

$$\begin{aligned} \mathcal{E}_n^2(c) &= \left(1 - \prod_{s=1}^{c-1} q_s\right)^2 \\ &\le 1 - \prod_{s=1}^{c-1} q_s \\ &\le \sum_{s=1}^{c-1} (1 - q_s) \\ &\le \sum_{s=1}^{c-1} s/m \\ &\le c^2/2m\ . \end{aligned}$$

Further, as above, use the following estimate

$$E[\mathrm{Per}A_n^c(\eta)]^2 = n^{[c]} .$$

It follows, from (4.45), that

$$\begin{aligned} E\Delta_{n1}^2 &= \sum_{c=m+1}^{n} \left(\frac{m}{nv}\right)^{2c} n^{[c]}/c! \\ &\le \sum_{c=m+1}^{n} \left(\frac{m}{v\sqrt{n}}\right)^{c} /c! \longrightarrow 0 \end{aligned}$$

and

$$\begin{aligned} E\Delta_{n2}^2 &\le \sum_{c=2}^{m} \left(\frac{m}{nv}\right)^{2c} \mathcal{E}_n^2(c) n^{[c]}/c! \\ &\le c/m \longrightarrow 0 \text{ , as } \quad m \to \infty. \end{aligned}$$

These last two estimates complete the proof of lemma 4.2.

Consider a multiple sample

$$\xi_{sj} \ , \quad 1 \le s \le k \ , \quad 1 \le j \le n \ ,$$

of random variables which are jointly independent and identically distributed for fixed s.

Theorem 4.8. *Assume that the following conditions hold*

$$E\xi_{sj} = \mu_s \neq 0,$$

$$E(\xi_{sj} - \mu_s)^2 = \sigma_s^2 > 0,$$

$$v_s := \mu_s/\sigma_s,$$

and

$$m_s/\sqrt{n} \to \gamma_s > 0 \ , \quad 1 \le s \le k \ , \quad \text{as } n \to \infty \tag{4.46}$$

Then the weak convergence

$$n^{-[m]} \prod_{s=1}^{k} \mu_s^{-m_s} \mathrm{Per}A_n^m(\xi_1, \ldots, \xi_k) \Rightarrow \exp\left[\sum_{s=1}^{k} (a_s \tau_s - a_s^2/2)\right] \tag{4.47}$$

takes place. Here,

$$\begin{aligned} a_s &:= \gamma_s/v_s \\ &= \gamma_s \sigma_s/\mu_s, \end{aligned}$$

and τ_s, $1 \leq s \leq k$, are standard normally distributed mutually independent random variables.

The proof of theorem 4.8 is presented at first for the case of $k = 2$. In the general case, $k \geq 2$, it is much more complicated and involves cumbersome calculations.

Introduce the normalized variables

$$\eta_{sj} := (\xi_{sj} - \mu_s)/\sigma_s,$$

$$\xi_{sj}/\mu_s = 1 + \eta_{sj}/v_s,$$

and consider the normalized permanent

$$\mathrm{Per}_n := n^{-[m]}\mu_1^{-m_1}\mu_2^{-m_2}\mathrm{Per}A_n^m(1+\eta_1/v_1,\ 1+\eta_2/v_2).$$

By using the decomposition formula (1.11), get

$$\mathrm{Per}_n = n^{-[m]}m_1! \sum_{q\in Q_{m_1}} \prod_{s=1}^{m_1}(1+\eta_{1q_s}/v_1)\ m_2! \sum_{q'\in Q_{m_2}^{n-m_1}(q)} \prod_{s=1}^{m_2}(1+\eta_{2q'_s}/v_2). \tag{4.48}$$

The inner sums in (4.48) are symmetric m_2-degree polynomials of $n - m_1$ variables η_{2j}, $1 \leq j \leq n$ with deleted variables η_{2q_s}, $1 \leq s \leq m_1$. Thus, employing the factor $m_2!$,

$$m_2! \sum_{q'\in Q_{m_2}^{n-m_1}(q)} \prod_{s=1}^{m_2}(1+\eta_{2q'_s}/v_2) = \mathrm{Per}A_{n-m_1}^{m_2}(1+\eta_2/v_2;q).$$

By the shift decomposition formula (1.22),

$$(n-m_1)^{-[m_2]}\mathrm{Per}A_{n-m_1}^{m_2}(1+\eta_2/v_2;q)$$

$$= \sum_{c=0}^{m_2} \frac{m_2^{[c]}}{(n-m)^{[c]}}\mathrm{Per}A_{n-m_1}^{c}(\eta_2/v_2;q)/c!.$$

According to the Cauchy formula for residues and the definition of the generating function (1.13) - (1.14), get the representation

$$\mathrm{Per}A_{n-m_1}^{c}(\eta_2/v_2;q)/c! = \frac{1}{2\pi i}\oint \frac{dz}{z^{c+1}}\prod_{j=1}^{n}(1+z\eta_{2j}/v_2)\prod_{s=1}^{m_1}(1+z\eta_{2q_2}/v_2)^{-1}.$$

Combining these results, change the order of sums in (4.48)

$$\mathrm{Per}_n = \frac{(n-m_1)^{[m_2]}}{n^{[m]}}\sum_{c=0}^{m_2}\frac{m_2^{[c]}}{(n-m_1)^{[c]}}\frac{1}{2\pi i}\oint\frac{dz}{z^{c+1}}$$

$$\times\prod_{j=1}^{n}(1+z\eta_{2j}/v_2)\ m_1!\sum_{q\in Q_{m_1}}\prod_{s=1}^{m_1}(1+\eta_{1q_s}/v_1)(1+z\eta_{2q_s}/v_2)^{-1}. \tag{4.49}$$

The product in (4.49) can be transformed as follows

$$(1+\eta_{1j}/v_1)(1+z\eta_{2j}/v_2)=1+\eta_{2j}(z),$$

where

$$\eta_{2j}(z):=(\eta_{1j}/v_1-z\eta_{2j}/v_2)(1+z\eta_{2j}/v_2)^{-1}. \tag{4.50}$$

Now the inner sum in (4.49), with the factor $m_1!$, is the permanent of a sampling matrix $A_n^{m_1}(1+\eta(z))$ which can be represented by the shift decomposition formula (1.22) as follows

$$m_1!\sum_{q\in Q_{m_1}}\prod_{s=1}^{m_1}(1+\eta_{q_s}(z))=n^{[m_1]}\sum_{d=0}^{m_1}\frac{m_1^{[d]}}{n^{[d]}}\mathrm{Per}A_n^d(\eta(z))/d!.$$

Inserting this representation into (4.49) and taking into account the fact that

$$(n-m_1)^{[m_2]}n^{[m_1]}/n^{[m]}=1\ ,$$

obtain the following formula

$$\mathrm{Per}_n=\sum_{c=0}^{m_2}\frac{m_2^{[c]}}{(n-m_1)^{[c]}}\frac{1}{2\pi i}\oint\frac{dz}{z^{c+1}}$$

$$\times\prod_{j=1}^{n}(1+z\eta_{2j}/v_2)\sum_{d=0}^{m_1}\frac{m_1^{[d]}}{n^{[d]}d!}\mathrm{Per}A_n^d(\eta(z)). \tag{4.51}$$

Investigate the sums in (4.51) by using condition (4.46). First employ the following asymptotic representation of the coefficient in the inner sum of (4.51)

$$m_1^{[d]}/n^{[d]}=(\gamma/\sqrt{n})^d[1+\mathcal{E}_n(d)],$$

where

$$\mathcal{E}_n(d)\longrightarrow 0\quad\text{as}\quad n\to\infty\ . \tag{4.52}$$

As in the proof of theorem 4.7,

$$\sum_{d=0}^{m_1}\frac{m_1^{[d]}}{n^{[d]}}\mathrm{Per}A_n^d(\eta(z))/d!$$

$$=\sum_{d=0}^{n}(\gamma_1/\sqrt{n})^d\mathrm{Per}A_n^d(\eta(z))/d!+\Delta_{n1}(z)+\Delta_{n2}(z),$$

where

$$\Delta_{n1}(z)=\sum_{d=0}^{m_1}\mathcal{E}_n(d)\mathrm{Per}A_n^d(\eta(z))/d!,$$

$$\Delta_{n2}(z)=\sum_{d=m_1+1}^{n}(\gamma_1/\sqrt{n})^d\mathrm{Per}A_n^d(\eta(z))/d!\ ,$$

and $\mathcal{E}_n(d)$ is defined in (3.8) by $b = \gamma_1/\sqrt{n}$. By virtue of estimate (3.17) and theorem 4.3,

$$\Delta_{nk}(z) \xrightarrow{P} 0 \quad \text{as } n \to \infty\,, \quad k = 1, 2\,. \tag{4.53}$$

By the definition of generating function (1.13) - (1.14),

$$\begin{aligned}\Pi_n(z) &= \sum_{d=0}^{n} (\gamma_1/\sqrt{n})^d \mathrm{Per} A_n^d(\eta(z))/d! \\ &= \prod_{j=1}^{n} (1 + \gamma_1 \eta_j(z)/\sqrt{n})\,.\end{aligned} \tag{4.54}$$

Taking into account (4.50), compute

$$1 + \gamma_1 \eta_j(z)/\sqrt{n}$$

$$= [1 + a_1 \eta_{1j}/\sqrt{n} + z(1 - \gamma_1/\sqrt{n})\eta_{2j}/v_2](1 + z\eta_{2j}/v_2)^{-1}.$$

Inserting this result into (4.54) and then into (4.51), obtain

$$\mathrm{Per}_n = \sum_{c=0}^{m_2} \frac{m_2^{[c]}}{(n - m_1)^{[c]}} \frac{1}{2\pi i} \oint \frac{dz}{z^{c+1}} \Pi_n(z) + \Delta_{n1} + \Delta_{n2}\,, \tag{4.55}$$

where

$$\Pi_n(z) = \prod_{j=1}^{n} [1 + a_1 \eta_{1j}/\sqrt{n} + z(1 - \gamma_1/\sqrt{n})\eta_{2j}/v_2]$$

and

$$\Delta_{nk} = \sum_{c=0}^{m_2} \frac{m_2^{[c]}}{(n - m_1)^{[c]}} \frac{1}{2\pi i} \oint \frac{dz}{z^{c+1}} \Delta_{nk}(z).$$

Changing the variable $z \to z/\sqrt{n}$ and employing (4.53),

$$\Delta_{nk} \xrightarrow{P} 0 \quad \text{as } n \to \infty\,, \quad k = 1, 2\,. \tag{4.56}$$

Now investigate the asymptotic behavior of the sum in (4.55). Under condition (4.46), the asymptotic behavior of the sum in (4.55) has the following form

$$m_2^{[c]}/(n - m_1)^{[c]} = \gamma_n^c(1 + \mathcal{E}_n(c)),$$

where $\gamma_n := \gamma_2/(\sqrt{n} - \gamma_1)$ and $\mathcal{E}_n(c)$ satisfies condition (4.52). Changing the order of the sum and the integral in (4.55), obtain

$$\begin{aligned}\mathrm{Per}_n &= \frac{1}{2\pi i} \oint \frac{dz}{z} \Pi_n(z) \sum_{c=0}^{m_2} (\gamma_n/z)^c + \Delta_n' \\ &= \frac{1}{2\pi i} \oint \frac{dz}{z - \gamma_n} \Pi_n(z)[1 - (\gamma_n/z)^{m_2+1}] + \Delta_n'\end{aligned}$$

with Δ'_n satisfying condition (4.56). By using the Cauchy integral formula,

$$\mathrm{Per}_n = \Pi_n(\gamma_n) + \Delta_n.$$

Taking into account (4.50),

$$\Pi_n(\gamma_n) = \prod_{j=1}^{n}(1 + a_1\eta_{1j}/\sqrt{n} + a_2\eta_{2j}/\sqrt{n}).$$

The completion of the proof of theorem 4.8 is given by means of theorem 4.2 according to which

$$\Pi_n(\gamma_n) \Rightarrow \exp\left(\sum_{s=1}^{2}(a_s\tau_s - a_s^2/2)\right) \quad \text{as } n \to \infty .$$

Prove theorem 4.8 in the general case of $k \geq 2$. By means of the decomposition formula (1.34)

$$n^{-[m]}\prod_{s=1}^{k}\mu_s^{-m_s}\mathrm{Per}A_n^m(\xi_1,\dots,\xi_k)$$

$$= \sum_{c=0}^{m} n^{-[c]} \sum_{r_1+\dots+r_k=c}\prod_{s=1}^{k}(\sigma_s/\mu_s)^{r_s}m_s^{[r_s]}\mathrm{Per}\ A_n^c(\eta_1,\dots,\eta_k)/\prod_{s=1}^{k}r_s! \tag{4.57}$$

and, in addition,

$$\eta_s = (\eta_{sj}\ , \quad 1 \leq j \leq n),$$

$$\eta_{sj} = \sigma_s^{-1}(\xi_{sj} - \mu_s),$$

$$E\eta_{sj} = 0\ , \quad E\eta_{sj}^2 = 1, \quad s = 1,\dots,k.$$

Represent (4.57) in the form

$$n^{-[m]}\prod_{s=1}^{k}\mu_s^{-m_s}\mathrm{Per}A_n^m(\xi_1,\dots,\xi_k) = \Pi_n + \Delta_{n1} + \Delta_{n2} - \Delta_{n3}\ , \tag{4.58}$$

where

$$\Pi_n = \sum_{c=0}^{n}\sum_{r_1+\dots+r_k=c}\prod_{s=1}^{k}\left(\frac{\sigma_s}{\mu_s}\frac{m_s}{n}\right)^{r_s}\mathrm{Per}A_n^c(\eta_1,\dots,\eta_k)/\prod_{s=1}^{k}r_s!,$$

$$\Delta_{n1} = \sum_{c=2}^{m}\sum_{r_1+\dots+r_k=c}(n^{-[c]} - n^{-c})\prod_{s=1}^{k}\left(\frac{\sigma_s}{\mu_s}\right)^{r_s}m_s^{[r_s]} \times$$

$$\times \quad \mathrm{Per}A_n^c(\eta_1,\dots,\eta_k)/\prod_{s=1}^{k}r_s!,$$

$$\Delta_{n2}=\sum_{c=2}^{m}n^{-c}\sum_{r_1+\cdots+r_k=c}\prod_{s=1}^{k}\left(\frac{\sigma_s}{\mu_s}\right)^{r_s}\left(\prod_{s=1}^{k}m_s^{[r_s]}-\prod_{s=1}^{k}m_s^{r_s}\right)\times$$
$$\times\mathrm{Per}\ A_n^c(\eta_1,\ldots,\eta_k)/\prod_{s=1}^{k}r_s!,$$

$$\Delta_{n3}=\sum_{c=m+1}^{n}\sum_{r_1+\cdots+r_k=c}\prod_{s=1}^{k}\left(\frac{\sigma_s}{\mu_s}\frac{m_s}{n}\right)^{r_s}$$
$$\times\mathrm{Per}\ A_n^c(\eta_1,\ldots,\eta_k)/\prod_{s=1}^{k}r_s!.$$

The first term in the right-hand side of (4.58) coincides with the generating multiple sample function. That is why, according to formula (1.15),

$$\Pi_n=\prod_{j=1}^{n}\left(1+\frac{1}{\sqrt{n}}\sum_{s=1}^{k}\left(\frac{\sigma_s}{\mu_s}\frac{m_s}{\sqrt{n}}\right)\eta_{sj}\right). \tag{4.59}$$

If $n\to\infty$, then (4.47) and (4.59) provide the weak convergence

$$\Pi_n\Rightarrow\exp\left(\sum_{s=1}^{k}(a_s\tau_s-a_s^2/2)\right). \tag{4.60}$$

Verify that the other terms in (4.58) satisfy the condition

$$\Delta_{ns}\xrightarrow{P}0,\ n\to\infty,1\le s\le 3. \tag{4.61}$$

First notice that

$$E|\mathrm{Per}\ A_n^c(\eta_1,\ldots,\eta_k)|\le n^{[c]/2} \tag{4.62}$$

according to the Hölder inequality.

The relation

$$\begin{aligned}E|\Delta_{n1}| &\le n^{-1}\prod_{s=1}^{m-1}(1-\frac{s}{n})^{-1}\sum_{c=2}^{m}c(c-1)\\ &\times\sum_{d_1+\cdots+d_k=c}\prod_{s=1}^{k}(\frac{\sigma_s}{\mu_s}\frac{m_s}{\sqrt{n}})^{d_s}/\prod_{s=1}^{k}d_s!\\ &= n^{-1}\prod_{s=1}^{m-1}(1-\frac{s}{n})^{-1}\sum_{c=2}^{m}\frac{1}{(c-2)!}\left(\sum_{s=1}^{k}(\frac{\sigma_s}{\mu_s}\frac{m_s}{\sqrt{n}})\right)^{c}\\ &\le n^{-1}\prod_{s=1}^{m-1}(1-\frac{s}{n})^{-1}\left(\sum_{s=1}^{k}(\frac{\sigma_s}{\mu_s}\frac{m_s}{\sqrt{n}})\right)^{2}\exp\left(\sum_{s=1}^{k}(\frac{\sigma_s}{\mu_s}\frac{m_s}{\sqrt{n}})\right)\end{aligned}$$

is true because

$$\begin{aligned} n^{-[c]} - n^{-c} &= n^{-c} \prod_{s=1}^{c-1} (1 - \frac{s}{n})^{-1} \; (1 - \prod_{s=1}^{c-1} (1 - \frac{s}{n})) \\ &\leq n^{-c} \prod_{s=1}^{m-1} (1 - \frac{s}{n})^{-1} \frac{c(c-1)}{2n} \end{aligned}$$

in Δ_{n1}, for all $2 \leq c \leq m$.
In this relation

$$\sum_{s=1}^{k} (\frac{\sigma_s}{\mu_s} \frac{m_s}{\sqrt{n}}) \to \sum_{s=1}^{k} a_s,$$

$$\prod_{s=1}^{m-1} (1 - \frac{s}{n})^{-1} \to \exp((\sum_{s=1}^{k} \gamma_s)^2/2) .$$

That is why $E|\Delta_{n1}| \to 0$, and consequently (4.61) when $s = 1$. In order to estimate Δ_{n2}

$$\begin{aligned} \prod_{s=1}^{k} m_s^{r_s} - \prod_{s=1}^{k} m_s^{[r_s]} &= \prod_{s=1}^{k} m_s^{r_s} \left(1 - \prod_{s=1}^{k} \prod_{i=1}^{r_s - 1} (1 - \frac{i}{m_s}) \right) \\ &\leq \prod_{s=1}^{k} m_s^{r_s} (\sum_{s=1}^{k} \frac{r_s(r_s - 1)}{2m_s}) \\ &\leq \prod_{s=1}^{k} m_s^{r_s} (c(c-1) \sum_{s=1}^{k} m_s^{-1}). \end{aligned}$$

Further, with (4.62),

$$\begin{aligned} E|\Delta_{n2}| &\leq (\sum_{s=1}^{k} m_s^{-1}) \sum_{c=2}^{m} c(c-1) \\ &\quad \times \sum_{r_1 + \cdots + r_k = c} \prod_{s=1}^{k} (\frac{\sigma_s}{\mu_s} \frac{m_s}{\sqrt{n}})^{r_s} / \prod_{s=1}^{k} r_s! \\ &= (\sum_{s=1}^{k} m_s^{-1}) \sum_{c=2}^{m} \frac{1}{(c-2)!} \left(\sum_{s=1}^{k} (\frac{\sigma_s}{\mu_s} \frac{m_s}{\sqrt{n}}) \right)^c \\ &\leq (\sum_{s=1}^{k} m_s^{-1}) \left(\sum_{s=1}^{k} (\frac{\sigma_s}{\mu_s} \frac{m_s}{\sqrt{n}}) \right)^2 \cdot \exp(\sum_{s=1}^{k} (\frac{\sigma_s}{\mu_s} \frac{m_s}{\sqrt{n}})). \end{aligned}$$

That is why $E|\Delta_{n2}| \to 0$ under the conditions of the theorem. In order to estimate Δ_{n3}, with the help of (4.62),

$$\begin{aligned} E|\Delta_{n_3}| &\le \sum_{c=m+1}^{n} \sum_{r_1+\cdots+r_k=c} \prod_{s=1}^{k} (\frac{\sigma_s}{\mu_s}\frac{m_s}{\sqrt{n}})^{r_s} / \prod_{s=1}^{k} r_s! \\ &= \sum_{c=m+1}^{n} \frac{1}{c!}\left(\sum_{s=1}^{k}(\frac{\sigma_s}{\mu_s}\frac{m_s}{\sqrt{n}})\right)^c \\ &\le \frac{1}{(m+1)!}\left(\frac{\sigma_s}{\mu_s}\frac{m_s}{\sqrt{n}}\right)^{m+1} \exp(\sum_{s=1}^{k}(\frac{\sigma_s}{\mu_s}\frac{m_s}{\sqrt{n}})). \end{aligned}$$

Consequently $E|\Delta_{n3}| \to 0$ and therefore (4.61) is true. As a result, the relations (4.58)-(4.60) provide (4.47) for all $k \ge 2$.

4.7 Functional limit theorems

Consider the multiple sampling matrix $A_n^m(\xi_1, \ldots, \xi_k)$ with independent vectors, where $\xi_{s1}, \xi_{s_2}, \ldots, \xi_{sn}$ are independent and identically distributed for every fixed s.

For $mn^{-1} \le t \le 1$, define a multiple sampling permanent random process

$$\text{Per } A_{[nt]}^m(\xi_1, \ldots, \xi_k), \tag{4.63}$$

where the symbol $[nt]$ denotes the integer part of nt. If $0 \le t \le mn^{-1}$. Then extend (4.63) in the following way:

$$\text{Per } A_{[nt]}^m(\xi_1, \ldots, \xi_k) = 0$$

with probability 1. The trajectories of the process (4.63) belong to the space $\mathcal{D}[0,1]$ i.e. the space of all real-valued functions without breaks of the second kind.

Theorem 4.9. *Under the conditions of theorem 4.4 for sampling permanent random process (4.63), the following weak convergence occurs*

$$\text{Per } A_{[nt]}^m(\xi/\sigma\sqrt{n}) \Rightarrow H_m(w(t), t) \tag{4.64}$$

in the topology of the space $\mathcal{D}[0,1]$, where $w(t)$ is the standard Wiener process.

Proof. Represent the permanent process by means of the Cauchy integral

$$\text{Per } A_{[nt]}^m(\xi/\sigma\sqrt{n}) = \frac{m!}{2\pi i}\oint \frac{dz}{z^{m+1}} \prod_{j=1}^{[nt]}(1 + \frac{z}{\sigma\sqrt{n}}\xi_j). \tag{4.65}$$

Hence, starting at (4.65), the functional variant of (4.33) is deduced:

$$\text{Per } A_{[nt]}^m(\xi/\sigma\sqrt{n}) = H_m\left(\frac{\sigma_{[nt],1}}{\sigma\sqrt{n}}, \frac{\sigma_{[nt],2}}{\sigma^2 n}\right)$$

$$+\sum_{k=3}^{m} P_k(n^{-3/2}\sigma_{[nt],3},\dots,n^{-k/2}\sigma_{[nt],k})$$
$$\times H_{m-k}(\frac{\sigma_{[nt],1}}{\sigma\sqrt{n}},\frac{\sigma_{[nt],2}}{\sigma^2 n})m^{[k]}, \tag{4.66}$$

where

$$\sigma_{[nt],k}=\sum_{j=1}^{[nt]}\xi_j^k,\ k=1,2,\dots .$$

According to the Donscker principle (see Billingsley (1968))

$$\sigma^{-1}n^{-1/2}\sigma_{[nt],1}\Rightarrow w(t)$$

and according to the strong law of large numbers

$$\sigma^{-2}n^{-1}\sigma_{[nt],2}\to t\quad (a.s.).$$

Hence

$$H_m\left(\frac{\sigma_{[nt],1}}{\sigma\sqrt{n}},\frac{\sigma_{[nt],2}}{\sigma^2 n}\right)\Rightarrow H_m(w(t),t) \tag{4.67}$$

in the space $\mathcal{D}[0,1]$ when $n\to\infty$.

Further it is obvious that, for all $k\geq 3$,

$$\sup_{0\leq t\leq 1}|\sigma_{[nt],k}|\leq\sum_{j=1}^{n}|\xi_j|^k,$$

and, according to the Marzinkevich–Kolmogorov law of large numbers,

$$n^{-k/2}\sum_{j=1}^{n}|\xi_j|^k\to 0\quad (a.s.).$$

By virtue of this in (4.66), and for $k\geq 3$,

$$\sup_{0\leq t\leq 1}|P_k(n^{-3/2}\sigma_{[nt],3},\dots,n^{-k/2}\sigma_{[nt],k})|\to 0\quad (a.s.). \tag{4.68}$$

From (4.66)–(4.68) (4.64) follows.

Theorem 4.10. *If* $\mu_s=0$ *in the conditions of theorem* 4.5, *then the following relation is true*

$$\text{Per }A_{[nt]}^m(\xi_1/\sigma_1\sqrt{n},\dots,\xi_k/\sigma_k\sqrt{n})\Rightarrow\prod_{s=1}^{k}H_{m_s}(w_s(t),t) \tag{4.69}$$

in the topological space $\mathcal{D}[0,1]$, *where* $w_1(t),\dots,w_k(t)$ *are independent standard Wiener processes.*

Proof. First a formula for process (4.63) analogous to (4.65) can be written

$$\text{Per } A^m_{[nt]}(\xi_1/\sigma_1\sqrt{n},\dots,\xi_k/\sigma_k\sqrt{n})$$

$$= \frac{m_1!\dots m_k!}{(2\pi i)^k}\oint\dots\oint\prod_{s=1}^{k}\frac{dz_s}{z_s^{m_s+1}}$$

$$\times\prod_{j=1}^{[nt]}(1+\frac{1}{\sqrt{n}}\sum_{s=1}^{k} z_s\sigma_s^{-1}\xi_{sj}). \tag{4.70}$$

Developing the expression in the right-hand side of (4.70) in the same way as in the proof of theorem 4.5, (4.69) is produced.

Theorem 4.11. *Let $\mu \neq 0$, $0 < \sigma^2 < \infty$ and condition $m/\sqrt{n} \to \gamma$, $\gamma > 0$. Then the weak convergence as $n \to \infty$ takes place*

$$[nt]^{-[m]}\text{Per } A^m_{[nt]}(\xi/\mu) \Rightarrow \exp\left\{-\frac{1}{2}\left(\frac{\sigma\gamma}{\mu}\right)^2\frac{1}{t}+\frac{\sigma\gamma}{\mu}\frac{w(t)}{t}\right\} \tag{4.71}$$

in the topology of space $\mathcal{D}(0,1]$.

Proof. The random process Per $A^m_{[nt]}(\xi/\mu)$ can be decomposed by the decomposition formula with a shift (1.32)

$$[nt]^{-[m]}\text{Per } A^m_{[nt]}(\xi/\mu)$$

$$= \sum_{r=0}^{m}\left(\frac{\sigma}{\mu}\right)^r m^{[r]}[nt]^{-[r]}\text{Per } A^r_{[nt]}((\xi-\mu)/\sigma)/r!. \tag{4.72}$$

According to theorem 4.9

$$n^{-r/2}\text{Per } A^r_{[nt]}((\xi-\mu)/\sigma) \Rightarrow H_r(w(t),t).$$

Besides this,

$$m^{[r]}[nt]^{-[r]}n^{r/2} \to \left(\frac{\gamma}{t}\right)^r,$$

when $0 < t \leq 1$. Therefore (4.72) provides

$$[nt]^{-[m]}\text{Per } A^m_{[nt]}(\xi/\mu) \Rightarrow \sum_{r=0}^{\infty}\left(\frac{\sigma\gamma}{\mu t}\right)^r H_r(w(t),t)/r!.$$

According to the definition of Hermite polynomials

$$\sum_{r=0}^{\infty}\left(\frac{\sigma\gamma}{\mu t}\right)^r H_r(w(t),t)/r!$$

$$= \exp\left\{-\frac{1}{2}\left(\frac{\sigma\gamma}{\mu}\right)^2\frac{1}{t}+\frac{\sigma\gamma}{\mu}\frac{w(t)}{t}\right\},$$

that proves (4.71).

Theorem 4.12. *If the conditions of theorem 4.8 are satisfied then*

$$[nt]^{-[m]}\text{Per } A_n^m(\xi_1/\mu_1, \dots, \xi_k/\mu_k)$$

$$\Rightarrow \exp\{-\frac{1}{2t}\sum_{s=1}^{k}\left(\frac{\sigma_s\gamma_s}{\mu_s}\right)^2 + \frac{1}{t}\sum_{s=1}^{k}\frac{\sigma_s\gamma_s}{\mu_s}w_s(t)\}$$

in topological space $\mathcal{D}(0,1]$, where $w_1(t), \dots, w_k(t)$ are independent standard Wiener processes.

Proof. It is sufficient to apply the method of the proof of previous theorems in this section.

Chapter 5

Symmetric statistics

Permanent symmetric statistics are considered as an integral over permanent random measures with deterministic kernels.

5.1 Permanent random measures

Let $\zeta_j(B)$, $1 \le j \le n$ be a family of orthogonal random measures on a measurable space $(X, \mathcal{X})$ which are mutually independent and identically distributed with the structure function

$$P(B \cap B') = E\zeta_j(B)\zeta_j(B'),\ \ B, B' \in \mathcal{X}.$$

Here $P(B)$ is a structure measure on $(X, \mathcal{X})$.

The permanent random measure on a measurable space $(X, \mathcal{X})$ is determined by the following form

$$\pi_n^{\zeta}(B_1, \ldots, B_m) := n^{-[m]}\mathrm{Per}A_n^m(\zeta(B)), \tag{5.1}$$

where

$$A_n^m(\zeta(B)) = [a_{ij} = \zeta_j(B_i); 1 \le i \le m\ , 1 \le j \le n],$$

or

$$\pi_n^{\zeta}(B_1, \ldots, B_m) = n^{-[m]} \sum_{1 \le j_1 \ne \ldots \ne j_m \le n} \prod_{s=1}^{m} \zeta_{j_s}(B_s)\ . \tag{5.2}$$

Here $n^{-[m]} := 1/n^{[m]} = 1/n(n-1)\ldots(n-m+1)$.

Example 5.1. The Dirac permanent measure is defined by a simple sample

$$\xi_j,\ 1 \le j \le n$$

of independent and identically distributed random variables taking values in a measurable space $(X, \mathcal{X})$ with the distribution function

$$P(B) := \mathcal{P}\{\xi_j \in B\}$$

of the following form (see section 2.2):

$$\pi_n^\delta(B_1, \dots, B_m) := n^{-[m]} \mathrm{Per} A_n^m(\delta_\xi(B)),$$

where

$$A_n^m(\delta_\xi(B)) := [a_{ij} = \delta_{\xi_j}(B_i); 1 \le i \le m, \quad 1 \le j \le n],$$

and

$$\delta_x(B) := \begin{cases} 1, & x \in B, \\ 0, & x \notin B, \end{cases} \qquad x \in X, \quad B \in \mathcal{X}$$

is the Dirac measure on $(X, \mathcal{X})$.

In the specific case of $m = 1$, the Dirac permanent measure

$$\pi_n^\delta(B) := n^{-1} \sum_{j=1}^n \delta_{\xi_j}(B)$$

is an empirical distribution of the sample ξ_j, $1 \le j \le n$.

Example 5.2. The Gaussian permanent measure is defined as follows

$$\pi_n^W(B_1, \dots, B_m) := n^{-[m]} \mathrm{Per} A_n^m(W(B)) \ ,$$

where

$$A_n^m(W(B)) := [a_{ij} = W_j(B_i); \ 1 \le i \le m, \ 1 \le j \le n] \tag{5.3}$$

and $W_j(B)$, $1 \le j \le n$, are Gaussian random measures which are mutually independent and identically distributed with zero mean and the covariance function

$$P(B \cap B') = E W_j(B) W_j(B').$$

The permanent of the matrix (5.3) can be represented by corollary 1.1 with $a_{sj} = W_j(B_s)$ in the following form

$$\mathrm{Per} A_n^m(W(B)) = \prod_{s=1}^m \sum_{j_s=1}^{n-s+1} W_{j_s}(B_s), \tag{5.4}$$

where, in each sum, $j_s \ne j_1, \dots, j_{s-1}, 2 \le s \le m$. By the finite divisible property of a Gaussian measure,

$$\sum_{j=1}^{n_s} W_j(B) \stackrel{d}{=} \sqrt{n_s} W(B) \tag{5.5}$$

and for the disjoint sets $B_1, \dots, B_m$,

$$\prod_{s=1}^m W(B_s) \stackrel{d}{=} n^{-m/2} \prod_{s=1}^m \sum_{j=1}^n W_j(B_s) \ , \tag{5.6}$$

where "$\stackrel{d}{=}$" means equality by distribution.

Lemma 5.1. *For the disjoint sets* $B_1, \dots, B_m$

$$\mathrm{Per}A_n^m(W(B)) = \prod_{s=1}^{m}\sum_{j=1}^{n} W_j(B_s) - \Delta_n(B_1, \dots, B_m) , \tag{5.7}$$

where

$$E\Delta_n^2(B_1, \dots, B_m) = (n^m - n^{[m]})\prod_{s=1}^{m} P(B_s) . \tag{5.8}$$

Proof. The random function $\Delta_n(B_1, \dots, B_n)$ can be represented in an explicit form, for example,

$$\Delta_n(B_1, B_2) = \sum_{j=1}^{n} W_j(B_1)W_j(B_2).$$

Meanwhile, it is easily to verify that

$$E[\Delta_n(B_1, \dots, B_m)\mathrm{Per}A_n^m(W(B))] = 0 .$$

So, by (5.6) and (5.7),

$$E[\mathrm{Per}A_n^m(W(B))]^2 = n^m E[\prod_{s=1}^{m} W(B_s)]^2 - E\Delta_n^2.$$

When $B_1, \dots, B_m$ are disjoint sets,

$$\begin{aligned} E[\mathrm{Per}A_n^m(W(B))]^2 &= n^{[m]}E[\prod_{s=1}^{m} W(B_s)]^2 \\ &= n^{[m]}\prod_{s=1}^{m} P(B_s). \end{aligned}$$

Comparing the last two representations of the second moment of the permanent, obtain (5.8).

Corollary 5.1. *For disjoint sets* $B_1, \dots, B_m$ *the Gaussian permanent measure has the following representation*

$$\pi_n^W(B_1, \dots, B_m) = n^{-[m]}\prod_{s=1}^{m}\sum_{j=1}^{n} W_j(B_s) - \delta_n(B_1, \dots, B_m) , \tag{5.9}$$

where

$$E\delta_n^2(B_1, \dots, B_m) = n^{-[m]}(n^m/n^{[m]} - 1)\prod_{s=1}^{m} P(B_s).$$

Corollary 5.2. *For the disjoint sets* $B_1, \ldots, B_m$,

$$\prod_{s=1}^{m} W(B_s) \stackrel{\mathrm{d}}{=} n^{m/2} \pi_n^W (B_1, \ldots, B_m) + \varepsilon_n(B_1, \ldots, B_m), \tag{5.10}$$

where the function $\varepsilon_n(B_1, \ldots, B_m)$ *possesses the property*

$$E\varepsilon_n^2(B_1, \ldots, B_m)$$

$$= 2(1 - \prod_{s=1}^{m-1} (1 - \frac{s}{n})^{1/2}) \prod_{s=1}^{m} P(B_s). \tag{5.11}$$

Example 5.3. The Poisson permanent measure is defined by the formula

$$\pi_n^{\nu}(B_1, \ldots, B_m) := n^{-[m]} \mathrm{Per} A_n^m(\nu(B)),$$

where

$$A_n^m(\nu(B)) := [a_{ij} = \nu_j(B_i); \quad 1 \le i \le m, \ 1 \le j \le n]$$

with the Poisson random measures $\nu_j(B)$, $1 \le j \le n$, $B \in \mathcal{X}$, which are mutually independent and identically distributed according to the Poisson law with the intensity measure $P(B) := E\nu_j(B)$, $1 \le j \le n$, $B \in \mathcal{X}$.

Note that, by the definition of the Poisson permanent measure,

$$E\pi_n^{\nu}(B_1, \ldots, B_m) = \prod_{s=1}^{m} P(B_s).$$

Let $\Phi : X^m \to R$ be a symmetric real-valued kernel of m variables ($m \le n$).

Permanent symmetric statistics of degree m are determined by the integral over the permanent random measure

$$U_{nm}^{\zeta}(\Phi) := \int_X \cdots \int_X \Phi(x_1, \ldots, x_m) \pi_n^{\zeta}(dx_1, \ldots, dx_m) \tag{5.12}$$

or, using representation (5.2) for a permanent random measure, the permanent symmetric statistics (5.12) can be represented in the following form

$$U_{nm}^{\zeta}(\Phi) := n^{-[m]} \sum_{1 \le j_1 \ne \ldots \ne j_m \le n} \Phi(\zeta_{j_1}, \ldots, \zeta_{j_m}) , \tag{5.13}$$

where

$$\Phi(\zeta_{j_1}, \ldots, \zeta_{j_m}) := \int_X \cdots \int_X \Phi(x_1, \ldots, x_m) \prod_{s=1}^{m} \zeta_{j_s}(dx_s) . \tag{5.14}$$

$$\begin{aligned}\widehat{\Phi} &:= EU^{\zeta}_{nm}(\Phi)\\ &= \int_X \cdots \int_X \Phi(x_1,\dots,x_m)\prod_{s=1}^{m}\widehat{P}(dx_s),\end{aligned}$$

where

$$\widehat{P}(B) = E\zeta_j(B), B \in \mathcal{X}.$$

Note that the arguments ζ_{j_s}, $1 \le j \le m$ in (5.14) are fictitious and they are real only in the case of the Dirac permanent measure

$$U^{\delta}_{nm}(\Phi) := n^{-[m]} \sum_{1\le j_1 \neq \dots \neq j_m \le n} \Phi(\xi_{j_1},\dots,\xi_{j_m}) , \tag{5.15}$$

defined by the given sample ξ_j, $1 \le j \le n$.

The symmetric statistics (5.15) are well-known U-statistics (see Borovskikh et al. (1994)).

Example 5.4. The permanent symmetric statistics with a primitive kernel

$$\Phi(x_1,\dots,x_m) := \prod_{s=1}^{m}\varphi(x_s)$$

have the following representation

$$U^{\zeta}_{nm}(\varphi) = n^{-[m]}m!\; P^m_n(\zeta),$$

where

$$P^m_n(\zeta) := \sum_{1\le j_1 < \dots < j_m \le n}\prod_{s=1}^{m}\zeta_{j_s}$$

is a symmetric m-degree polynomial of the sample

$$\zeta_j := \int_X \varphi(x)\zeta_j(dx), \quad 1 \le j \le n.$$

Example 5.5. Let $\varphi_s : X \to R$, $1 \le s \le k$, be a real-valued measurable function on $(X, \mathcal{X})$. Construct a multiple sample as follows:

$$\zeta_{sj} := \int_X \varphi_s(x)\zeta_j(dx), \ 1 \le s \le k, \ 1 \le j \le n.$$

The multiple sampling symmetric statistic is defined by the formula

$$U^{\zeta}_{nm}(\varphi_1,\dots,\varphi_k) := n^{-[m]}\mathrm{Per}A^m_n(\zeta_1,\dots,\zeta_k) , \tag{5.16}$$

where

$$A_n^m(\zeta_1,\ldots,\zeta_k) := \begin{bmatrix} A_n^{m_1}(\zeta_1) \\ \cdots \\ A_n^{m_k}(\zeta_k) \end{bmatrix}$$

is a multiple sampling matrix with blocks

$$A_n^{m_s}(\zeta_s) := [a_{ij} = \zeta_{sj};\ 1 \le i \le m_s,\ 1 \le j \le n],\quad 1 \le s \le k\ ,$$

which are sampling matrices. Clearly the multiple sampling symmetric statistics (5.16) are determined by the degenerate kernel

$$\Phi_k(x_1,\ldots,x_m) := \prod_{s=1}^{k}\prod_{e=1}^{m_s} \varphi_s(x_{se}),$$

where

$$x_{se} := x_{\bar{m}_s+e},\ \bar{m}_s := m_1 + \ldots + m_{s-1},\quad s = 1,\ldots,k,$$

$$\bar{m}_1 = 0,\quad \bar{m}_k = m_1 + \cdots + m_{k-1}.$$

5.2 Normal and Poisson approximation scheme

Permanent symmetric statistics with fixed degree m over the permanent random measure (5.1)

$$U_{nm}^{\zeta}(\Phi) := \int_X \cdots \int_X \Phi(x_1,\ldots,x_m)\pi_n^{\zeta}(dx_1,\ldots,dx_m) \tag{5.17}$$

are considered as a statistical image of multiple stochastic integrals

$$I_m^W(\Phi) := \int_X \cdots \int_X \Phi(x_1,\ldots,x_m)\prod_{s=1}^{m} W(dx_s)$$

over the Wiener random measure $W(dx)$ or

$$I_m^{\bar{\nu}}(\Phi) := \int_X \cdots \int_X \Phi(x_1,\ldots,x_m)\prod_{s=1}^{m} \bar{\nu}(dx_s)$$

over the Poisson centered random measure $\bar{\nu}(dx) := \nu(dx) - \lambda(dx)$ with intensity $\lambda(dx) = E\nu(dx)$.

Theorem 5.1. *Let the real-valued symmetric kernel $\Phi : X^m \to R$ be square integrable*

$$\int_X \cdots \int_X \Phi^2(x_1,\ldots,x_m)\prod_{s=1}^{m} P(dx_s) < \infty$$

and completely degenerate

$$\int_X \Phi(x_1,\ldots,x_m)P(dx_1) = 0 \quad (a.s.)$$

Then the normalized permanent symmetric statistics $n^{[m]/2}U^{\zeta}_{nm}(\Phi)$ *are weakly convergent as* $n \to \infty$ *to the Ito stochastic integral* $I^W_m(\Phi)$*:*

$$n^{[m]/2}U^{\zeta}_{nm}(\Phi) \Rightarrow I^W_m(\Phi) \quad \text{as} \quad n \to \infty. \tag{5.18}$$

Proof. Consider the orthogonal system of real-valued functions $\varphi_r(x)$, $r \geq 0$, $\varphi_0(x) \equiv 1$ on the Hilbert space $L^2(X,P)$:

$$\int_X \varphi_r(x)\varphi_{r'}(x)P(dx) = \begin{cases} 1\,, & r = r', \\ 0\,, & r \neq r', \end{cases}$$

$$\int_X \varphi_r(x)P(dx) = 0\,, \quad r \geq 1.$$

The orthogonal system of symmetric statistics with degenerate kernels on the Hilbert space $L^2(\zeta)$ of random variables

$$\zeta := \int_X \varphi(x)\zeta(dx)\,, \quad \varphi \in L^2(X,P)$$

is determined in the form of multiple sampling symmetric statistics (see example 5.5)

$$U^{\zeta}_{nm}(\varphi_{r_1},\ldots,\varphi_{r_k}) := n^{-[m]/2}\mathrm{Per}A^m_n(\zeta_{r_1},\ldots,\zeta_{r_k}), \quad 1 \leq k \leq m$$

$$r := (r_1,\ldots,r_k)\,, \quad r_1 \neq \ldots \neq r_k \geq 1,$$

$$\zeta_{r_s} := (\zeta_{r_s j} := \int_X \varphi_{r_s}(x)\zeta_j(dx), \quad 1 \leq j \leq n)\,, 1 \leq s \leq k.$$

The orthogonal property of these statistics

$$E[U^{\zeta}_{nm}(\varphi_{r_1},\ldots,\varphi_{r_k})U^{\zeta}_{nm}(\varphi_{r'_1},\ldots,\varphi_{r'_k})] = \begin{cases} \prod_{s=1}^m m_s! & , \quad r = r', \\ 0 & , \quad r \neq r', \end{cases}$$

is a consequence of the orthogonal property of functions $\varphi_r(x), r \geq 0$.

Consider the Fourier series for symmetric statistics

$$n^{[m]/2}U^{\zeta}_{nm}(\Phi) = \sum_{k=1}^{m} \sum_{r_1\geq 1,\ldots,r_k\geq 1} \Phi_{kr}U^{\zeta}_{nm}(\varphi_{r_1},\ldots,\varphi_{r_k}) \tag{5.19}$$

with the Fourier coefficients Φ_{kr} determined by the relation

$$\Phi_{kr} = E[n^{[m]/2}U^{\zeta}_{nm}(\Phi)U^{\zeta}_{nm}(\varphi_{r_1},\ldots,\varphi_{r_k})]\,.$$

It can be verified that

$$\Phi_{kr} = \int_X \cdots \int_X \Phi(x_1, \ldots, x_m) \text{Per} A_m^m(\varphi_{r_1}, \ldots, \varphi_{r_k}) \prod_{s=1}^m P(dx_s)/m!,$$

where

$$A_m^m(\varphi_{r_1}, \ldots, \varphi_{r_k}) := \begin{bmatrix} A_m^{m_1}(\varphi_{r_1}) \\ \cdots \\ A_m^{m_k}(\varphi_{r_k}) \end{bmatrix}$$

and

$$A_m^{m_s}(\varphi_{r_s}) := [a_{ij} = \varphi_{r_s}(x_j); \ \ 1 \le i \le m_s \ , \ \ 1 \le j \le m],$$

$$1 \le s \le k, m = m_1 + \ldots + m_k.$$

Therefore the coefficients Φ_{kr} in the Fourier series expansion (5.19) are the Fourier coefficients of the deterministic kernel $\Phi(x_1, \ldots, x_m)$ over the orthogonal system of functions

$$\text{Per} A_m^m(\varphi_{r_1}, \ldots, \varphi_{r_k}) \ , \quad r_1, \ldots, r_k \ge 1, \ 1 \le k \le m \ .$$

on the Hilbert space $L^2(X^m, \ P^m)$.

Now, using the theorem 4.5, obtain the weak convergence

$$U_{nm}^{\varsigma}(\varphi_{r_1}, \ldots, \varphi_{r_k}) \Rightarrow \prod_{s=1}^k H_{m_s}(\tau_{r_s}) \quad \text{as} \quad n \to \infty \ .$$

Hence the weak convergence of the truncated series

$$U_{n,N}^{\varsigma}(\Phi) := \sum_{k=1}^m \sum_{1 \le r \le N} \Phi_{kr} U_{nm}^{\varsigma}(\varphi_{r_1}, \ldots, \varphi_{r_k})$$

$$\Rightarrow \sum_{k=1}^m \sum_{1 \le r \le N} \Phi_{kr} \prod_{s=1}^m H_{m_s}(\tau_{r_s}) \quad \text{as} \quad n \to \infty$$

takes place.

Note that the Ito–Wiener stochastic integral $I_m^W(\Phi)$ allows the following expansion (see Ito (1951), Ito et al. (1988))

$$I_m^W(\Phi) = \sum_{k=1}^m \sum_{r_1 \ge 1, \ldots, r_k \ge 1} \Phi_{kr} \prod_{s=1}^k H_{m_s}(\tau_{r_s}).$$

The standard technique for convergence in the mean square of the Fourier series gives statement (5.18) of theorem 5.1.

The analogous result for the Poisson stochastic integral $I_m^{\bar{\nu}}(\Phi)$ is more simple if considering the symmetric statistics over the permanent δ-measure (example 5.1)

$$U_{nm}^{\delta}(\Phi) := \int_X \cdots \int_X \Phi(x_1, \ldots, x_m) \pi_n^{\delta}(dx_1, \ldots, dx_m). \tag{5.20}$$

According to remark 2.2 and theorem 2.3, the symmetric statistic (5.20) has the following representation

$$n^{[m]}U^{\delta}_{nm}(\Phi) := \int_X \cdots \int_X \Phi(x_1, \ldots, x_m) \prod_{s=1}^{m} \nu_n(dx_s)$$

where

$$\nu_n(dx) := \sum_{j=1}^{n} \delta_{\xi_j}(dx),$$

$$\prod_{s=1}^{m} \nu_n(dx_s) = \prod_{s=1}^{m} \sum_{j_s=1}^{n-s+1} \delta_{j_s}(dx_s),$$

where in each sum $j_s \neq j_1, \ldots, j_{s-1}$, $2 \leq s \leq m$.

In the particular case of the degenerate kernel

$$\Phi(x_1, \ldots, x_m) = \prod_{s=1}^{m} \varphi(x_s),$$

as was mentioned in remark 2.5 (cf. formula (2.50)), under Poisson approximation condition (2.16) the weak convergence, as $n \to \infty$,

$$\begin{aligned} n^{[m]}U^{\delta}_{nm}(\varphi) \; &:= \; \int_{X^m} \prod_{s=1}^{m} \varphi(x_s)\nu_n(dx_s) \\ &\Rightarrow \; I^{\nu}_m(\varphi) \end{aligned}$$

takes place.

Due to the fact that for any finite continuous function $\Phi(x_1, \ldots, x_m)$ under condition (2.16),

$$n^m \int_{X^m} \Phi(x_1, \ldots, x_m) \prod_{s=1}^{m} P_n(dx_s) \to \int_{X^m} \Phi(x_1, \ldots, x_m) \prod_{s=1}^{m} \lambda(dx_s)$$

conclude that the symmetric statistics (5.20) under Poisson approximation condition (2.16) can be considered as the statistical image of the Poisson stochastic integrals $I^{\nu}_m(\Phi)$ and $I^{\bar{\nu}}_m(\Phi)$.

It is well-known (see Ito (1988)) that for the Poisson stochastic integrals $I^{\bar{\nu}}_m(\Phi)$ there exists the expansion over the Charlier stochastic polynomials:

$$I^{\bar{\nu}}_m(\Phi) = \sum_{k=1}^{m} \sum_{r \geq 1} \Phi_{kr} I^{\bar{\nu}}_m(\varphi_{r_1}, \ldots, \varphi_{r_k}) \tag{5.21}$$

where $r := (r_1, \ldots, r_k)$, $r_1 \neq \cdots \neq r_k$ are the multi-indices, $\{\varphi_r, r \geq 0, \varphi_0 \equiv 1\}$ is the orthogonal system of real-valued functions on the Hilbert space $L^2(X, \lambda)$ and Φ_{kr} are the Fourier coefficients of the function Φ determined by the relation

$$\Phi_{kr} = \int_{X^m} \Phi(x_1, \ldots, x_m) \mathrm{Per}\; A^m_m(\varphi_{r_1}, \ldots, \varphi_{r_k}) \prod_{s=1}^{m} \lambda(dx_s)/m!.$$

Analogously, permanent symmetric statistic (5.20) permits the decomposition

$$n^{[m]}U^{\delta}_{nm}(\Phi) = \sum_{k=1}^{m}\sum_{r\geq 1}\Phi_{kr}U^{\delta}_{nm}(\bar{\varphi}_{r_1},\dots,\bar{\varphi}_{r_k}) \tag{5.22}$$

by symmetric statistics with degenerate kernels

$$U^{\delta}_{nm}(\bar{\varphi}_{r_1},\dots,\bar{\varphi}_{r_k}) := \text{Per } A^m_n(\bar{\varphi}_{r_1},\dots,\bar{\varphi}_{r_k}).$$

$I^{\bar{\nu}}_m(\varphi_{r_1},\dots,\varphi_{r_k})$ and Per $A^m_n(\bar{\varphi}_{r_1},\dots,\bar{\varphi}_{r_k})$ in explicit form are given in remark 2.7.

Thus relations (2.69) and (2.73) are compatible with decompositions (5.21) and (5.22) and anticipate the following theorem.

Theorem 5.2. *Let $\Phi : X^m \to R$ be a symmetric continuous function such that there exist a compact set $B \in X$ and the constant $c > 0$ and*

$$\Phi(x_1,\dots,x_m) = 0, \quad \textit{if} \quad (x_1,\dots,x_m) \notin B^m$$

and

$$|\Phi(x_1,\dots,x_m)| \leq c^m, \quad \textit{if} \quad (x_1,\dots,x_m) \in B^m.$$

Then under the conditions of Poisson approximation and when m is fixed, the weak convergences

$$n^{[m]}U^{\delta}_{nm}(\Phi) \Rightarrow I^{\nu}_m(\Phi)$$

and

$$n^{[m]}U^{\bar{\delta}}_{nm}(\Phi) \Rightarrow I^{\bar{\nu}}_m(\Phi)$$

take place.

The proof of theorem 5.2 is given in the next section (5.3) which is devoted to the demonstration of weak convergence for symmetric statistics both in Gaussian and Poisson schemes and in the more general scheme of infinitely divisible approximation.

5.3 Composition method

In order to prove weak convergence of functionals of type (5.12) extend the classical composition method (Sazonov (1981)).

Principal identity of the composition method.

Let F and G be arbitrary probability measures on a measurable space $(X, \mathcal{X})$, and F^n and G^n be their n-multiple convolutions.

The main idea of the composition method consists of the identity

$$F^n - G^n = \sum_{k=1}^{n} F^{n-k}G^{k-1}(F - G) \tag{5.23}$$

where the derivative is understood in the sense of the convolution of distributions. Each term in the right-hand side of (5.23) can be written as

$$F^{n-k}G^{n-k}(F-G)$$

$$=(F^{n-k}-G^{n-k})G^{k-1}(F-G)+G^{n-1}(F-G), k=1,\dots n,$$

and then identity (5.23) can be applied to the difference $F^{n-k}-G^{n-k}$. Formula (5.23) will be written by means of characteristic functions. Let

$$S_n = X_1 + \cdots + X_n,$$

and

$$T_n = Y_1 + \cdots + Y_n,$$

where X_i has the distribution F and Y_i has the distribution G.

Consider the characteristic functions

$$\varphi_n(t) = E\exp(itS_n),$$

$$\psi_n(t) = E\exp(itT_n), t \in R.$$

The formula

$$\varphi_n(t) - \psi_n(t) = \sum_{k=1}^{n} E(e^{itz_{nk}}(e^{itX_k} - e^{itY_k})) \tag{5.24}$$

is analogous to (5.23) for the difference $\varphi_n(t) - \psi_n(t)$, where

$$z_{nk} = \sum_{j=1}^{k-1} Y_j + \sum_{j=k+1}^{n} X_j.$$

Formula (5.24) allows estimation of the proximity of the characteristic functions $\varphi_n(t)$ and $\psi_n(t)$ in terms of proximity of the moments of the individual terms X_k and Y_k in sums S_n and T_n.

Adjust this method to comparison of the distributions and characteristic functions of permanent symmetric statistics of type (5.12). Explain in detail the permanent variant of the composition method, taking as examples Gaussian random measure and Poisson random measure. Further this method, without special additional reasoning, can be extended to a wide class of infinitely divisible random measures.

Weak convergence to the stochastic Wiener–Ito integral.

First introduce the following notation. For all $k = 0, 1, \dots, n$ define the matrices of $m \times n$ size

$$A_{nk}^{m} = [a_{ij}^{k};\ 1 \le i \le m,\ 1 \le j \le n]$$

with elements

$$a_{ij}^k = \begin{cases} W_j(dx_i)(n-i+1)^{-1/2} & , \quad 1 \le i \le m,\ 1 \le j \le k, \\ (\delta_{\xi_j}(dx_i) - P(dx_i))(n-i+1)^{-1/2} & , \quad 1 \le i \le m, k+1 \le j \le n. \end{cases}$$

Consider the matrices A_{nk}^m when $k = 0$,

$$A_{n0}^m = [a_{ij}^0 = (\delta_{\xi_j}(dx_i) - P(dx_i))(n-i+1)^{-1/2};\ 1 \le i \le m, 1 \le j \le n]$$

and A_{nk}^m when $k = n$,

$$A_{nn}^m = [a_{ij}^n = W_j(dx_i)(n-i+1)^{-1/2};\ 1 \le i \le m, 1 \le j \le n].$$

It is obvious that

$$\text{Per } A_{nk}^m = \text{Per } A_{nk}^m(dx_1, \ldots, dx_m)$$

is a permanent random measure when $k = 0, 1, \ldots, n$. Regarding this measure introduce the following functionals

$$U_k^{\delta W}(\Phi) = \int_{X^m} \Phi(x_1, \ldots, x_m) \text{Per } A_{nk}^m(dx_1, \ldots, dx_m),$$

where the symmetric function Φ is degenerate

$$\int_X \Phi(x_1, \ldots, x_{m-1}, y) P(dy) = 0 \quad (a.s.).$$

If $k = 0$ then

$$U_0^{\delta W}(\Phi) = n^{[m]/2} U_{nm}^{\delta}(\Phi),$$

and $U_{nm}^{\delta}(\Phi)$ is represented through the permanent measure in (5.12).

If $k = n$, then

$$U_n^{\delta W}(\Phi) := n^{[m]/2} \int_{X^m} \Phi(x_1, \ldots, x_m) \pi_n^W(dx_1, \ldots, dx_m). \tag{5.25}$$

Introduce the characteristic functions

$$\varphi_j(t) = E\exp(itU_j^{\delta W}(\Phi)), t \in R,$$

for all $j = 0, 1, \ldots, n$.

Lemma 5.2. *Let $\Phi \in L^2$ satisfy the condition of complete degeneration. Then, for every fixed $t \in R$,*

$$\varphi_0(t) - \varphi_n(t) \to 0, \tag{5.26}$$

when $n \to \infty$.

Proof. First write an analogy to (5.24)

$$\varphi_0(t) - \varphi_n(t) = \sum_{j=1}^{n} (\varphi_{j-1}(t) - \varphi_j(t)). \tag{5.27}$$

Estimate the differences $\varphi_{j-1}(t) - \varphi_j(t)$ from the right-hand side in (5.27). For this purpose introduce matrices $A_{nk}^m(-|k)$ of $m \times (n-1)$ dimension, i.e.

$$A_{nk}^m(-|k) = [a_{ij}^k; 1 \le i \le m, 1 \le j \le n, j \ne k]$$

with elements

$$a_{ij}^k = \begin{cases} W_j(dx_i)(n-i+1)^{-1/2}, & 1 \le i \le m, \ 1 \le j \le k-1, \\ (\delta_{\xi_j}(dx_i) - P(dx_i))(n-i+1)^{-1/2}, & 1 \le i \le m, \ k+1 \le j \le n, \end{cases}$$

and matrices of $n \times n$ dimension

$$EA_{nj}^m = \begin{pmatrix} E \\ A_{nj}^m \end{pmatrix},$$

where E is the matrix of $(n-m) \times n$ dimension composed of units.

Apply the Laplace formula for permanents (1.8). Then

$$\text{Per } EA_{n,j-1}^m = (n-m)!\text{Per } A_{n,j-1}^m,$$

$$\text{Per } EA_{nj}^m = (n-m)!\text{Per } A_{nj}^m.$$

According to formula (1.6) for matrix $EA_{n,j-1}^m$,

$$\begin{aligned} \text{Per } EA_{n,j-1}^m &= \sum_{s=1}^{m} \delta_{\xi_j}(dx_s)(n-s+1)^{-1/2}\text{Per } EA_{n,j-1}^m(s|j) \\ &+ \sum_{s=m+1}^{n} \text{Per } EA_{n,j-1}(s|j). \end{aligned}$$

According to the same formula for matrix EA_{nj}^m,

$$\begin{aligned} \text{Per } EA_{nj}^m &= \sum_{s=1}^{m} W_j(dx_s)(n-s+1)^{-1/2}\text{Per } EA_{nj}^m(s|j) \\ &+ \sum_{s=m+1}^{n} \text{Per } EA_{nj}^m(s|j). \end{aligned}$$

If $1 \le s \le m$, then

$$\begin{aligned} \text{Per } EA_{n,j-1}^m(s|j) &= \text{Per } EA_{nj}^m(s|j) \\ &= (n-m)!\text{Per } A_{nj}^m(s|j). \end{aligned}$$

If $m+1 \le s \le n$, then

$$\begin{aligned}\text{Per } EA^m_{n,j-1}(s|j) &= \text{Per } EA^m_{nj}(s|j) \\ &= (n-1-m)!\text{Per } A^m_{nj}(-|j).\end{aligned}$$

Consequently for all $j = 1, 2, \ldots, n$

$$\begin{aligned}\text{Per } A^m_{n,j-1} &= \text{Per } A^m_{nj}(-|j) \\ &+ \sum_{s=1}^{m} (\delta_{\xi_j}(dx_s) - P(dx_s))(n-s+1)^{-1/2}\text{Per } A^m_{nj}(s|j),\end{aligned}$$

$$\begin{aligned}\text{Per } A^m_{nj} &= \text{Per } A^m_{nj}(-|j) \\ &+ \sum_{s=1}^{m} W_j(dx_s)(n-s+1)^{-1/2}\text{Per } A^m_{nj}(s|j).\end{aligned}$$

Taking this fact into account

$$U^{\delta W}_{j-1}(\Phi) = U^{\delta W}_{n_j}(\Phi) + R_j + \bar{R}_j, \tag{5.28}$$

where

$$U^{\delta W}_{nj}(\Phi) = \int_{X^m} \Phi(x_1, \ldots, x_m)\text{Per } A^m_{nj}(-|j)(dx_1, \ldots, dx_m),$$

$$R_j = \sum_{s=1}^{m} (n-s+1)^{-1/2} r_{js}(\xi_j),$$

$$\bar{R}_j = \sum_{s=1}^{m} (n-s+1)^{-1/2} \bar{r}_{js}(\xi_j).$$

Moreover

$$\begin{aligned}r_{js}(x) &= \int_X \cdots \int_X \Phi(x_1, \ldots, x_m) I(|\Phi| \le \sqrt{n}) \\ &\times (\delta_x(dx_s) - P(dx_s))\text{Per } A^m_{nj}(s|j)(dx_1, \ldots, dx_{s-1}, dx_{s+1}, \ldots dx_m),\end{aligned}$$

$$\begin{aligned}\bar{r}_{js}(x) &= \int_X \cdots \int_X \Phi(x_1, \ldots, x_m) I(|\Phi| > \sqrt{n}) \\ &\times (\delta_x(dx_s) - P(dx_s))\text{Per } A^m_{nj}(s|j)(dx_1, \ldots, dx_{s-1}, dx_{s+1}, \ldots dx_m).\end{aligned}$$

In the analogous way

$$U_j^{\delta W}(\Phi) = U_{nj}^{\delta W}(\Phi) + D_j + \bar{D}_j, \tag{5.29}$$

where

$$D_j = \sum_{s=1}^{m} (n-s+1)^{-1/2} \int_X r_{js}(x) W_j(dx),$$

$$\bar{D}_j = \sum_{s=1}^{m} (n-s+1)^{-1/2} \int_X \bar{r}_{js}(x) W_j(dx).$$

Demonstrate that

$$r_{ks}(x) = (n-s+1)^{-1/2} \sum_{c=0}^{m-1} (m-1)^{[c]}$$

$$\times \sum_{1 \le \beta_1 \ne \cdots \ne \beta_c \le k-1} \int_X \cdots \int_X f_c(y_1, \ldots, y_c; \xi) \prod_{r=1}^{c} W_{\beta_r}(dy_r)/c!, \tag{5.30}$$

where

$$f_c(y_1, \ldots, y_c; \xi) \; = \sum_{k+1 \le \beta_{c+1} \ne \cdots \ne \beta_{m-1} \le n} \int_X \cdots \int_X \Phi(y_1, \ldots, y_c, y_{c+1}, \ldots, y_{m-1}, x)$$

$$\times I(|\Phi| \le \sqrt{n}) \prod_{r=c+1}^{m-1} (\delta_{\xi_{\beta_r}}(dy_r) - P(dy_r)).$$

For this purpose represent matrix $A_{nk}^m(s|k)$ as the sum

$$A_{nk}^m(s|k) = (n-s+1)^{1/2} n^{-[m]/2} (A+B),$$

where A is the matrix of $(m-1) \times (n-1)$ dimension with elements

$$a_{ij} = \begin{cases} W_j(dy_i), & 1 \le i \le m-1, \;\; 1 \le j \le k-1, \\ 0, & 1 \le i \le m-1, \;\; k+1 \le j \le n, \end{cases}$$

and B is the matrix of $(m-1) \times (n-1)$ dimension with elements

$$b_{ij} = \begin{cases} 0, & 1 \le i \le m-1, \;\; 1 \le j \le k-1, \\ \delta_{\xi_j}(dy_i), & 1 \le i \le m-1, \;\; k+1 \le j \le n, \end{cases}$$

and

$$(y_1, \ldots, y_{m-1}) = (x_1, \ldots, x_{s-1}, x_{s+1}, \ldots, x_m).$$

Note that

$$\text{Per } A_{nk}^m(s|k)(dx_1, \ldots, dx_{s-1}, dx_{s+1}, \ldots, dx_m)$$

$$= (n-s+1)^{1/2} n^{-[m]/2} \text{Per } (A+B)(dy_1, \ldots, dy_{m-1}).$$

Apply (1.45) to the permanent of the sum $A+B$. According to this formula and taking into account the specific character of matrices A and B,

$$\text{Per }(A+B)=\sum_{c=0}^{m-1}\sum_{\alpha\in Q_{c,m-1}}\sum_{\beta\in Q_{c,k-1}}\text{Per }A[\alpha|\beta]\text{Per }B(\alpha|\beta), \tag{5.31}$$

where, for the selected matrices A and B,

$$\text{Per }A[\alpha|\beta]=\sum_{(j_1,\dots,j_c)}\prod_{r=1}^{c}W_{\beta_{j_r}}(dy_{\alpha_r}),$$

$$\text{Per }B(\alpha|\beta)=\sum_{k+1\le\beta_{c+1}\ne\dots\ne\beta_{m-1}\le n}\prod_{r=c+1}^{m-1}(\delta_{\xi_{\beta_r}}(dy_{\alpha'_r})-P(dy_{\alpha'_r}))$$

where $\alpha'=(\alpha'_{c+1},\dots,\alpha'_{m-1})$ is a consequence complementary to α in $(1,2,\dots,m-1)$.

Considering (5.31) it can be seen that

$$\text{Per }A^m_{nk}(s|k)(dx_1,\dots,dx_{s-1},dx_{s+1},\dots,dx_m)$$

$$=\sum_{c=0}^{m-1}\sum_{1\le\alpha_1<\dots<\alpha_c\le m-1}\sum_{1\le\beta_1\ne\dots\ne\beta_c\le k-1}\prod_{r=1}^{c}W_{\beta_r}(dy_{\alpha_r})$$

$$\times\sum_{k+1\le\beta_{c+1}\ne\dots\ne\beta_{m-1}\le n}\prod_{r=c+1}^{m-1}(\delta_{\xi_{\beta_r}}(dy_{\alpha'_r})-P((dy_{\alpha'_r}))). \tag{5.32}$$

With the substitution of (5.32) in the definition of $r_{ks}(x)$ obtain formula (5.30).

Formula (5.30) is true for $\bar r_{ks}(x)$ with only the distinction that instead of an indicator function $I(|\Phi|\le\sqrt{n})$ it should be the indicator function $I(|\Phi|>\sqrt{n})$.

Returning to (5.27) consider the expression of the left-hand side. First, with the help of (5.28) and (5.29),

$$\varphi_{j-1}(t)=Ee^{itU^{\delta W}_{nj}(\Phi)}E_{\xi_j}e^{it(R_j+\bar R_j)},$$

$$\varphi_j(t)=Ee^{itU^{\delta W}_{nj}(\Phi)}E_{W_j}e^{it(D_j+\bar D_j)},$$

where E_{ξ_j} and E_{W_j} denote the conditional expectations relative to the random variables ξ_j and W_j. With the substitution of these formulas in (5.27),

$$\varphi_0(t)-\varphi_n(t)=$$

$$=\sum_{j=1}^{n}Ee^{itU^{\delta W}_{nj}(\Phi)}\left(E_{\xi_j}e^{it(R_j+\bar R_j)}-E_{W_j}e^{it(D_j+\bar D_j)}\right). \tag{5.33}$$

Consequently obtain an inequality

$$|\varphi_0(t)-\varphi_n(t)|\le\sum_{j=1}^{n}E|\Delta_j(t)|, \tag{5.34}$$

where

$$\Delta_j(t) = E_{\xi_j} e^{it(R_j+\bar{R}_j)} - E_{W_j} e^{it(D_j+\bar{D}_j)}.$$

In order to estimate $\Delta_j(t)$ apply the formula

$$e^{iz} = \sum_{s=0}^{r-1} (iz)^s \frac{1}{s!} + (iz)^r \frac{1}{(r-1)!} \int_0^1 (1-y)^{r-1} e^{izy} dy, \tag{5.35}$$

which is valid for all complex z and $r = 1, 2, \ldots$

With the choice of $r = 2$ and $z = t\bar{R}_j$, and $z = t\bar{D}_j$, obtain

$$\begin{aligned}
\Delta_j(t) &= E_{\xi_j} e^{itR_j} - E_{W_j} e^{itD_j} \\
&+ it(E_{\xi_j}(\bar{R}_{ij} e^{itR_j}) - E_{W_j}(\bar{D}_j e^{itD_j})) \\
&- t^2 \int_0^1 dy (1-y) E_{\xi_j}(\bar{R}_j^2 e^{it(R_j+\bar{R}_j y)}) \\
&+ t^2 \int_0^1 dy (1-y) E_{W_j}(\bar{D}_j^2 e^{it(D_j+\bar{D}_j y)}).
\end{aligned} \tag{5.36}$$

Further, in order to estimate the first difference of the right-hand side of (5.36), choose $r = 3$ in (5.35). Then

$$\begin{aligned}
E_{\xi_j} e^{itR_j} - E_{W_j} e^{itD_j} &= it(E_{\xi_j} R_j - E_{W_j} D_j) \\
&- \frac{t^2}{2}(E_{\xi_j} R_j^2 - E_{W_j} D_j^2) \\
&+ \frac{(it)^3}{2} \int_0^1 dy (1-y)^2 E_{\xi_j}(R_j^3 e^{ityR_j}) \\
&- \frac{(it)^3}{2} \int_0^1 dy (1-y)^2 E_{W_j}(D_j^3 e^{ityD_j}).
\end{aligned} \tag{5.37}$$

In (5.37),

$$\begin{aligned}
E_{\xi_j} R_j &= E_{W_j} D_j = 0, \\
E_{\xi_j} R_j^2 &= E_{W_j} D_j^2 \\
&= \int_X \left(\sum_{s=1}^{m} (n-s+1)^{-1/2} r_{js}(x) \right)^2 P(dx).
\end{aligned}$$

In order to estimate the second difference in (5.36), choose $r = 1$. Then

$$E_{\xi_j}(\bar{R}_j e^{itR_j}) - E_{W_j}(\bar{D}_j e^{itD_j})$$

$$= E_{\xi_j}\bar{R}_j - E_{W_j}\bar{D}_j + it\int_0^1 dy E_{\xi_j}(\bar{R}_j R_j e^{itR_j y})$$

$$-it\int_0^1 dy E_{W_j}(\bar{D}_j D_j e^{itD_j y}). \tag{5.38}$$

Here

$$E_{\xi_j} R_j = E_{W_j} D_j = 0.$$

Therefore, from (5.37) and (5.38), obtain

$$|E_{\xi_j} e^{itR_j} - E_{W_j} e^{itD_j}| \le \frac{|t^3|}{6}(E_{\xi_j}|R_j|^3 + E_{W_j}|D_j|^3),$$

$$|E_{\xi_j}(\bar{R}_j e^{itR_j}) - E_{W_j}(\bar{D}_j)e^{itD_j}| \le |t|(E_{\xi_j}|\bar{R}_j R_j| + E_{W_j}|\bar{D}_j D_j|). \tag{5.39}$$

In order to estimate the other terms in the right-hand side of (5.36), apply the inequalities

$$|E_{\xi_j}(\bar{R}_j e^{itR_j})| \le E_{\xi_j}|\bar{R}_j|,$$

$$|E_{W_j}(\bar{D}_j e^{itD_j}) \le E_{W_j}|\bar{D}_j|,$$

$$|E_{\xi_j}(\bar{R}_j^2 e^{it(R_j+\bar{R}_j y)})| \le E_{\xi_j}\bar{R}_j^2,$$

$$|E_{W_j}(\bar{D}_j^2 e^{it(D_j+\bar{D}_j y)})| \le E_{W_j}\bar{D}_j^2. \tag{5.40}$$

With the substitution of (5.37) in (5.40) and (5.36) it is seen that

$$\begin{aligned} |\Delta_j(t)| \le\ & t^2(E_{\xi_j}|\bar{R}_j R_j| + E_{W_j}|\bar{D}_j D_j|) \\ & + \frac{t^2}{2}(E_{\xi_j}\bar{R}_j^2 + E_{W_j}\bar{D}_j^2) \\ & + \frac{|t|^3}{6}(E_{\xi_j}|R_j|^3 + E_{W_j}|D_j|^3). \end{aligned} \tag{5.41}$$

In (5.41),

$$E_{\xi_j}\bar{R}_j^2 = E_{W_j}\bar{D}_j^2, \tag{5.42}$$

$$E_{\xi_j}|\bar{R}_j R_j| \le (E_{\xi_j}\bar{R}_j^2)^{1/2}(E_{W_j}R_j^2)^{1/2},$$

$$E_{W_j}|\bar{D}_j D_j| \le (E_{W_j}\bar{D}_j^2)^{1/2}(E_{W_j}D_j^2)^{1/2}$$
$$= (E_{\xi_j}\bar{R}_j^2)^{1/2}(E_{\xi_j}\bar{R}_j^2)^{1/2}.$$

Moreover

$$\sum_{j=1}^{n}(E_{\xi_j}\bar{R}_j^2)^{1/2}(E_{\xi_j}R_j^2)^{1/2}$$
$$\le \left(\sum_{j=1}^{n} E_{\xi_j}\bar{R}_j^2\right)^{1/2}\left(\sum_{j=1}^{n} E_{\xi_j}R_j^2\right)^{1/2}. \qquad (5.43)$$

Show that

$$E_{W_j}|D_j|^3 \le 2E_{\xi_j}|R_j|^3. \qquad (5.44)$$

Write, according to the definition,

$$E_{W_j}|D_j|^3 = E_{W_j}|\int_X R_j(x)W_j(dx)|^3$$
$$= E|\tau|^3(E_{\xi_j}R_j^2)^{3/2}$$
$$\le E|\tau|^3 E_{\xi_j}|R_j|^3,$$

where τ is a standard normal random variable and

$$E|\tau|^3 = (8\pi)^{1/2} \le 2.$$

Substitute (5.42) and (5.44) in (5.41). Then, taking into account (5.43), obtain

$$|\varphi_0(t) - \varphi_n(t)| \le 2t^2(\sum_{j=1}^{n} E\bar{R}^2)^{1/2}(\sum_{j=1}^{n} ER_j^2)^{1/2}$$
$$+ t^2\sum_{j=1}^{n} E\bar{R}_j^2 + \frac{1}{2}|t|^3 \sum_{j=1}^{n} E|R_j|^3. \qquad (5.45)$$

Consider the expectation in the right-hand side of (5.45).

The estimate of ER_k^2. Apply (5.30). According to this formula

$$ER_k^2 = m^2 n^{-[m]} \sum_{c=0}^{m-1}(m-1)^{2[c]}(k-1)^{[c]}$$
$$\times \int_X \cdots \int_X Ef_c^2(y_1, \ldots, y_c; \xi) \prod_{r=1}^{c} P(dy_r)/(c!)^2. \qquad (5.46)$$

The random variable $f_c(y_1,\dots,y_c;\xi)$ is adduced in (5.30) in explicit form and its structure coincides with the structure of U-statistics. That is why, according to the Hoeffding formula for variance of U-statistics (see Borovskikh et al. (1994)),

$$Ef_c^2(y_1,\dots,y_c;\xi)=(n-k)^{[m-c-1]}Eg_{nc}^2(y_1,\dots,y_c;\ \xi_{c+1},\dots,\xi_m) \tag{5.47}$$

where

$$g_{nc}(y_1,\dots,y_m)=\int_X\cdots\int_X\Phi(y_1,\dots,y_c,x_{c+1},\dots,x_m)I(|\Phi|\le\sqrt{n})$$
$$\times\prod_{s=c+1}^{m}(\delta_{y_s}(dx_s)-P(dx_s)).$$

Considering (5.46)-(5.47) observe that

$$ER_k^2=m^2n^{-[m]}\sum_{c=0}^{m-1}(m-1)^{2[c]}(k-1)^{[c]}(n-k)^{[m-c-1]}Eg_{nc}^2/(c!)^2. \tag{5.48}$$

From (5.48) obtain the estimate

$$\sum_{k=1}^{n}ER_k^2\le m^2n^mn^{-[m]}\sum_{c=0}^{m-1}(m-1)^{2[c]}Eg_{nc}^2/(c!)^2 \tag{5.49}$$

because

$$(k-1)^{[c]}(n-k)^{[m-c-1]}\le n^{m-1}.$$

Estimate of $E\bar{R}_k^2$.

It is obvious that formula (5.48) is valid for $E\bar{R}_k^2$ as well but the function $\bar{g}_{nc}$ should be substituted instead of g_{nc} where

$$\bar{g}_{nc}(y_1,\dots,y_m)=$$

$$=\int_X\cdots\int_X\Phi(y_1,\dots,y_c,x_{c+1},\dots,x_m)I(|\phi|>\sqrt{n})\prod_{s=c+1}^{m}(\delta_{y_s}(dx_s)-P(dx_s)).$$

That is why the following estimate is true:

$$\sum_{k=1}^{n}ER_k^2\le m^2n^mn^{-[m]}\sum_{c=0}^{m-1}(m-1)^{2[c]}E\bar{g}_{nc}^2/(c!)^2, \tag{5.50}$$

where

$$E\bar{g}_{nc}^2\le a(m)E(\Phi^2I(|\Phi|>\sqrt{n})),$$

and the constant $a(m)$ is dependent on m only.

Estimate of $E|R_k|^3$.

Apply (5.30) again. Taking into account this formula and inequalities for integral Wiener–Ito moments

$$E|R_k|^3 \le n^{-3[m]/2} \sum_{c=0}^{m-1} b_{1c}(m)(k-1)^{3[c]/2} \times \int_X \cdots \int_X E|f_c(y_1, \dots, y_c; x)|^3 \prod_{r=1}^{c} P(dy_r). \tag{5.51}$$

Further, according to the inequalities for U-statistics moments (see Borovskikh et al. (1994)),

$$E|f_c(y_1, \dots, y_c; \xi)|^3 \le b_{2c}(m)(n-k)^{3[m-c-1]/2} \times E|g_{nc}(y_1, \dots, y_c, \xi_{c+1}, \dots, \xi_m)|^3.$$

From (5.51) observe that

$$\sum_{k=1}^{n} E|R_k|^3 \le n^{-1/2} n^{3m/2} n^{-3[m]/2} \times \sum_{c=0}^{m-1} b_{1c}(m) b_{2c}(m) E|g_{nc}|^3 \tag{5.52}$$

where

$$E|g_{nc}|^3 \le b(m) \cdot E(|\Phi|^3 I(|\Phi| \le \sqrt{n}))$$

with the constant $b(m)$ dependent only on m. It follows from estimates (5.50) and (5.52) that

$$\sum_{k=1}^{n} E\bar{R}_k^2 \to 0,$$

$$\sum_{k=1}^{n} E|R|_k^3 \to 0$$

because

$$E(\Phi^2 I(|\Phi| > \sqrt{n})) \to 0,$$

$$n^{-1/2} E(|\Phi|^3 \cdot I(|\Phi| \le \sqrt{n})) \to 0$$

when $E\Phi^2 < \infty$.

These relations, together with (5.45), demonstrate that statement (5.26) is true.

Denote by

$$\varphi(t) := E\exp(itI_m^W(\Phi)), \quad t \in R.$$

Lemma 5.3. *If $\Phi \in L^2$ then*

$$\varphi_n(t) - \varphi(t) \to 0 \tag{5.53}$$

as $n \to \infty$, for every fixed $t \in R$.

Proof. According to (5.10) and (5.25)

$$I_m^W(\Phi) \stackrel{d}{=} U_n^{\delta W}(\Phi) + \alpha_n,$$

where

$$\alpha_n = \int_{X^m} \Phi(x_1, \dots, x_m)\varepsilon_n(dx_1, \dots, dx_m).$$

Taking into account the estimation (5.11),

$$E\alpha_n^2 = 2(1 - \prod_{s=1}^{m-1}(1 - \frac{s}{n})^{1/2}) \int_{X^m} \Phi^2(x_1, \dots x_m) \prod_{s=1}^{m} P(dx_s).$$

Further

$$\varphi_n(t) - \varphi(t) = E\{\exp(itU_n^{\delta W}(\Phi))(1 - \exp(it\alpha_n))\}.$$

Consequently

$$\begin{aligned} |\varphi_n(t) - \varphi(t)| &\leq |t|E|\alpha_n| \\ &\leq |t|(E\alpha_n^2)^{1/2}, \end{aligned}$$

that proves (5.53).

Lemma 5.4. *Under the conditions of lemma* 5.2

$$\varphi_0(t) - \varphi(t) \to 0, \tag{5.54}$$

as $n \to \infty$.

Proof. It is sufficient to remark that

$$|\varphi_0(t) - \varphi(t)| \leq |\varphi_0(t) - \varphi_n(t)| + |\varphi_n(t) - \varphi(t)|$$

and to apply (5.26) and (5.53).

The weak convergence

$$n^{[m]/2}U^{\delta}_{nm}(\Phi) \Rightarrow I^W_m(\Phi)$$

takes place, i.e. statement (5.18) of theorem 5.1 is true.

Weak convergence to the stochastic Ito–Poisson integral.

Let $\nu(B)$ be a Poisson random measure with intensity $E\nu(B) = \lambda(B)$, $B \in \mathcal{X}$ and $\bar{\nu}(B) = \nu(B) - \lambda(B)$ be a centered Poisson measure. Then $E\bar{\nu}(B) = 0$ and $E\bar{\nu}^2(B) = \lambda(B)$. The structural function of measure $\bar{\nu}(B)$ coincides with intensity $\lambda(B)$.

To estimate the stochastic Ito–Poisson integrals $I^{\nu}_m(\Phi)$ and $I^{\bar{\nu}}_m(\Phi)$ apply random sampling in a series scheme $\xi^{(n)}_j$, $1 \le j \le n$, with distribution $P_n(B) = P(\xi^{(n)}_j \in B)$, $B \in \mathcal{X}$, such that

$$nP_n(B) = \lambda(B), \quad B \in \mathcal{X},$$

i.e. suppose that the Poisson approximation scheme is valid.

Proof of theorem 5.2. For determination prove the second statement of the theorem about convergence to $I^{\bar{\nu}}_m(\Phi)$. Conversely this statement takes place if, for every fixed $t \in R$ and $n \to \infty$,

$$E\exp(itI^{\bar{\nu}}_m(\Phi)) - E\exp(itn^{[m]}U^{\delta}_{nm}(\Phi)) \to 0. \tag{5.55}$$

Prove relation (5.55) by the composition method repeating only key statements of the reasoning which brought (5.54).

According to the property of infinite divisibility of $\nu(B)$

$$\nu(B) \stackrel{\mathrm{d}}{=} (\nu^{(1)}(B) + \cdots + \nu^{(n)}(B))/n, B \in \mathcal{X},$$

where $\nu^{(1)}(B), \ldots, \nu^{(n)}(B)$ are independent copies of $\nu(B)$.

In view of this

$$I^{\bar{\nu}}_m(\Phi) \stackrel{d}{=} \int_{X^m} \Phi(x_1, \ldots, x_m) \prod_{s=1}^{m} (n^{-1} \sum_{j=1}^{n} (\nu^{(j)}(dx_s) - \lambda(dx_s))).$$

Further, for $k = 0, 1, \ldots, n$, introduce into consideration the matrices

$$A^m_{nk} = [a^k_{ij};\ 1 \le i \le m, 1 \le j \le n]$$

with elements

$$a^k_{ij} = \begin{cases} (\nu^{(j)}(dx_i) - \lambda(dx_i))/n, & 1 \le i \le m, \quad 1 \le j \le k, \\ \delta^{(n)}_j(dx_i) - \lambda(dx_i)/n, & 1 \le i \le m, \quad k+1 \le j \le n, \end{cases}$$

and the integral functionals with respect to permanent measures

$$\text{Per } A^m_{nk} = \text{Per } A_k(dx_1, \ldots, dx_m):$$

$$U^{\delta\bar{\nu}}_k(\Phi) = \int_{X^m} \Phi(x_1, \ldots, x_m) \text{Per } A_k(dx_1, \ldots, dx_m).$$

According to formula (1.49)

$$\text{Per } A^m_{nk}(dx_1,\ldots,dx_m) = \prod_{s=1}^{m}\left(n^{-1}\sum_{j=1}^{n}(\nu^{(j)}(dx_s)-\lambda(dx_s))\right) + \Delta_n(dx_1,\ldots,dx_m).$$

Therefore

$$\begin{aligned} U_k^{\bar\delta\bar\nu}(\Phi) &= \int_{X^m}\Phi(x_1,\ldots,x_m)\text{Per } A^m_{nk}(dx_1,\ldots,dx_m) \\ &= \int_{X^m}\Phi(x_1,\ldots,x_m)\prod_{s=1}^{m}\left(n^{-1}\sum_{j=1}^{n}(\nu^{(j)}(dx_s)-\lambda(dx_s))\right) \\ &+ \int_{X^m}\Phi(x_1,\ldots,x_m)\cdot\Delta_n(dx_1,\ldots,dx_m). \end{aligned}$$

Moreover

$$E\left|\int_{X^m}\Phi(x_1,\ldots,x_m)\Delta(dx_1,\ldots,dx_m)\right| \le c_1(\lambda)\cdot n^{-1}.$$

The estimate

$$|\varphi_0(t)-\varphi_n(t)| \le c_2(\lambda)t^2n^{-1}$$

is an analogy of (5.26) under the Poisson approximation condition, and the estimate

$$|\varphi_n(t)-\varphi(t)| \le c_3(\lambda)|t|n^{-1}$$

is an analogy of (5.53) for all $t \in R$. That is why, under this condition, (5.54) is possible to write as

$$|Ee^{itn^{[m]}U^{\delta}_{nm}(\Phi)} - Ee^{itI^{\nu}_m(\Phi)}| \le c_4(\lambda)(|t|+t^2)n^{-1}.$$

From this estimate (5.55) follows.

5.4 Approximation scheme with increasing degrees

In this section the Hoeffding decomposition for U-statistics is employed:

$$U^{\delta}_{nm}(\Phi) = \widehat{\Phi} + \sum_{c=r}^{m} m^{[c]}U^{\delta}_{nc}(g_{mc})/c! \tag{5.56}$$

where $r \ge 1$ is a rank of U-statistics and $g_{mc} := g_{mc}(x_1,\ldots,x_c)$ are completely degenerate kernels determined by the formula

$$g_{mc}(x_1,\ldots,x_c) := \int_{X^m} \Phi(y_1,\ldots,y_m)\prod_{s=1}^{c}(\delta_{x_s}(dy_s) - P(dy_s)) \prod_{s=c+1}^{m} P(dy_s).$$

First consider the situation

$$\gamma_n := m/\sqrt{n} \to 0, \quad \text{as } \ m,n \to \infty.$$

Example 5.6. Consider a primitive kernel

$$\Phi_m := \prod_{s=1}^{m} \varphi(x_s) \tag{5.57}$$

where $\varphi : X \to R$ is a function belonging to $L^2(X,P)$:

$$\int_X \varphi^2(x)P(dx) < +\infty\ .$$

It is easy to compute that

$$g_{mc} = \mu^{m-c}\prod_{s=1}^{c}(\varphi(x_s) - \mu)\ , \quad 1 \le c \le m,$$

where $\mu := \int_X \varphi(x)P(dx)$. Under the condition $\mu \neq 0$,

$$\begin{aligned} \mu^{-m} g_{mc} &= \mu^{-c}\prod_{s=1}^{c}(\varphi(x_s) - \mu) \\ &=: g_c. \end{aligned}$$

The function g_c does not depend on m. Consider the U-statistic (5.56) with the primitive kernel (5.57) on the sample ξ_j, $1 \le j \le n$, with the distribution function $P(B) := \mathcal{P}\{\xi_j \in B\}$.

Note that

$$\begin{aligned} Eg_c^2 &= Eg_c^2(\xi_1,\ldots,\xi_c) \\ &= (\sigma/\mu)^{2c} \\ &= v^{-2c} \end{aligned} \tag{5.58}$$

and

$$\begin{aligned} \widehat{\Phi} &= E\Phi_m \\ &= E\prod_{s=1}^{m}\varphi(\xi_s) \\ &= \mu^m \end{aligned}$$

where $\sigma^2 = E(\xi_1 - \mu)^2$.

Choosing the normalized coefficient, the following representation for U-statistics is obtained

$$[\mu^{-m+1}U^{\delta}_{nm}(\Phi_m) - \mu]/\sigma\gamma_n = \sum_{j=1}^{n}[\varphi(\xi_j) - \mu]/\sigma\sqrt{n} + \Delta_n ,$$

where

$$\gamma_n := m/\sqrt{n} \to 0, \quad \text{as} \quad m, n \to \infty,$$

by the assumption and

$$\Delta_n := \sum_{c=2}^{m} m^{[c]} U^{\delta}_{n}(g_c) v/\gamma_n .$$

Taking into account (5.58), obtain

$$E[U^{\delta}_{nc}(g_c)]^2 = c!\, Eg_c^2/n^{[c]},$$

and

$$\begin{aligned} E\Delta_n^2 &= \sum_{c=2}^{m}(m^{[c]})^2/n^{[c]}\gamma_n^2 v^{2(c-1)} \\ &\leq \sum_{c=2}^{m}\gamma_n^{2(c-1)}/v^{2(c-1)} \\ &\leq \gamma_n^2/(v^2 - \gamma_n^2) \to 0 \quad \text{as} \quad n \to \infty. \end{aligned}$$

By virtue of the central limit theorem, the following asymptotic result is obtained

$$[\mu^{-m+1}U^{\delta}_{nm}(\Phi_m) - \mu]/\sigma\gamma_n \Rightarrow \tau \quad \text{as} \quad m, n \to \infty.$$

Consider the general case.

Theorem 5.3. *Assume that the following conditions are satisfied:*

$$\gamma_n := m/\sqrt{n} \to 0 \quad \text{as} \quad m, n \to \infty;$$

$$E\Phi^2 := \int_{X^m} \Phi^2(x_1, \ldots, x_m) \prod_{s=1}^{m} P(dx_s) < \infty;$$

there exist normalized coefficients a_{mr} such that

$$\lim_{m \to \infty} a_{mr} g_{mc}(x_1, \ldots, x_c) = g_c(x_1, \ldots, x_c)$$

and

$$E[g_{mr} - g_r]^2 a_{mr}^2 \to 0 \quad \text{as} \quad m \to \infty, \tag{5.59}$$

$$a_{mr}^2 \sum_{c=r+1}^{m} \gamma_n^{2(c-r)} E g_{mc}^2 / c! \longrightarrow 0 \quad \text{as} \quad m \to \infty. \tag{5.60}$$

Then the weak convergence

$$a_{mr}[U_{nm}^{\delta}(\Phi) - \widehat{\Phi}]\gamma_n^{-r} \Rightarrow I_r^W(g_r)/r! \tag{5.61}$$

takes place. Here, $I_r^W(g_r)$ is the Ito stochastic integral over the Wiener random measure $W(B)$, $B \in \mathcal{X}$.

Proof. By using the Hoeffding decomposition formula (5.56), obtain the representation

$$a_{mr}[U_{nm}^{\delta}(\Phi) - \widehat{\Phi}]\gamma_n^{-r} = m^{[r]} U_{nr}^{\delta}(a_{mr} g_{mr})/\gamma_n^r r! + \Delta_n, \tag{5.62}$$

where

$$\Delta_n := \sum_{c=r+1}^{m} m^{[c]} U_{nc}^{\delta}(a_{mr} g_{mc})/\gamma_n^r c!$$

The first term in (5.62) can be transformed as follows

$$m^{[r]} U_{nr}^{\delta}(a_{mr} g_{mr})/\gamma_n^r r!$$

$$= n^{r/2} U_{nr}^{\delta}(g_r)/r! + n^{r/2} U_{nr}^{\delta}(g_r)(m^{[r]}/m^r - 1)/r! + \Delta_{nr}, \tag{5.63}$$

where

$$\Delta_{nr} := m^{[r]} U_{nr}^{\delta}(a_{mr}(g_{mr} - g_r))/\gamma_n^r r!$$

By virtue of theorem 5.1,

$$n^{r/2} U_{nr}^{\delta}(g_r) \Rightarrow I_r^W(g_r).$$

The second term in (5.63) tends to zero as $m \to \infty$. The last term in (5.63) allows the following estimate, under condition (5.59):

$$\begin{aligned} E\Delta_{nr}^2 &= (m^{[r]})^2 E[a_{mr}^2 (g_{mr} - g_r)^2]/n^{[r]} \gamma_n^{2r} r! \\ &\le E(g_{mr} - g_r)^2 a_{mr}^2 \to 0 \quad \text{as} \quad m \to \infty. \end{aligned}$$

Finally, condition (5.60) gives the opportunity to verify that the second term in (5.63) has the estimate

$$E\Delta_n^2 \le a_{mr}^2 \sum_{c=r+1}^{m} \gamma_n^{2(c-r)} E g_{mc}^2 \to 0.$$

The proof of theorem 5.3 can be employed in the situation, when $\gamma_n = m/\sqrt{n}$ has a finite limit as $m, n \to \infty$.

Theorem 5.4. *Assume that the following conditions are satisfied:*

$$\gamma_n = m/\sqrt{n} \longrightarrow \gamma \quad \text{as} \quad m \to \infty,$$

$$E\Phi^2 = \int_{X^m} \Phi^2(x_1, \ldots, x_m) \prod_{s=1}^{m} P(dx_s) < \infty,$$

there exist normalized coefficients a_{mr} such that

$$\lim_{m\to\infty} a_{mr} g_{mc}(x_1, \ldots, x_c) = g_c(x_1, \ldots, x_c),$$

$$\sum_{c=r}^{m} \gamma_n^{2c} E[a_{mr} g_{mr} - g_c]^2/c! \longrightarrow 0 \quad \text{as} \quad m \to \infty, \tag{5.64}$$

the series

$$\sum_{c=r}^{\infty} \gamma^{2c} E g_c^2/c! < \infty \tag{5.65}$$

is convergent.

Then the weak convergence takes place

$$a_{mr}[U_{nm}^{\delta}(\Phi) - \widehat{\Phi}] \Rightarrow \sum_{c=r}^{\infty} \gamma^c I_c^W(g_c)/c!\,. \tag{5.66}$$

Proof. First note that

$$m^{[c]} = \gamma_n^c n^{c/2} q_m(c),$$

where

$$q_m(c) := \prod_{s=1}^{c} (1 - s/m) \leq 1$$

and

$$\begin{aligned} [1 - q_m(c)]^2 &\leq 1 - q_m(c) \\ &\leq c^2/2m. \end{aligned}$$

Consider the normalized U-statistics under the Hoeffding decomposition (5.56)

$$a_{mr}[U_{nm}^{\delta}(\Phi) - \widehat{\Phi}] = \sum_{c=r}^{m} m^{[c]} U_{nc}^{\delta}(a_{mr} g_{mc})/c!\,. \tag{5.67}$$

The terms in (5.67) can be transformed as

$$m^{[c]}U^{\delta}_{nc}(a_{mr}g_{mc}) = \gamma^c_n n^{c/2}U^{\delta}_{nc}(g_c) + \Delta_n(c),$$

where

$$\Delta_n(c) := \gamma^c_n(1 - q_m(c))n^{c/2}U^{\delta}_{nc}(g_c) + \gamma^c_n q_m(c)n^{c/2}U^{\delta}_{nc}(a_{mr}g_{mc} - g_c). \tag{5.68}$$

Under condition (5.65), by virtue of theorem 5.1,

$$\sum_{c=r}^{m} \gamma^c_n n^{c/2}U^{\delta}_{nc}(g_c)/c! \Rightarrow \sum_{c=r}^{\infty} \gamma^c I^W_c(g_c)/c!.$$

Estimate the first sum in (5.68)

$$\begin{aligned} E\Delta^2_{n1}(c) &:= \sum_{c=r}^{m} \gamma^{2c}_n(1 - q_m(c))^2 n^c E[U^{\delta}_{nc}(g_c)]^2/(c!)^2 \\ &\leq \sum_{c=r}^{m} \gamma^{2c}_n(1 - q_m(c))^2 Eg^2_c/c! \\ &\leq m^{-1}\sum_{c=r}^{m^p} \gamma^{2c}_n c^2 Eg^2_c/c! + \sum_{c=m^p}^{m} \gamma^{2c}_n Eg^2_c/c! \\ &\leq m^{-1+2p}\sum_{c=r}^{m^p} \gamma^{2c}_n Eg^2_c/c! + \sum_{c=m^p}^{m} \gamma^{2c}_n Eg^2_c/c! \to 0, \end{aligned}$$

where $0 < p < 1/2$.
Further estimate the second sum in (5.68), under condition (5.64):

$$\begin{aligned} E\Delta^2_{n2}(c) &:= \sum_{c=r}^{m} \gamma^{2c}_n q^2_m(c) n^c E(U^{\delta}_{nc}(a_{mr}g_{mc} - g_c))^2/(c!)^2 \\ &\leq \sum_{c=r}^{m} \gamma^{2c}_n q^2_m(c) E(a_{mr}g_{mc} - g_c)^2/c! \longrightarrow 0. \end{aligned}$$

At the end of this section in the scheme of a series a simple sample,

$$\xi^{(n)}_j, \quad 1 \leq j \leq n, \tag{5.69}$$

of independent and identically distributed random variables taking values in a measurable space $(X, \mathcal{X})$ and having the distribution function $P_n(B) := \mathcal{P}\{\xi^{(n)}_j \in B\}$, $B \in \mathcal{X}$ is considered.

As in section 1.2, δ-measures are defined as

$$\begin{aligned} \delta^{(n)}_j(B) &:= I(\xi^{(n)}_j \in B) \\ &:= \begin{cases} 1, & \xi^{(n)}_j \in B, \\ 0, & \xi^{(n)}_j \notin B, \end{cases} \end{aligned}$$

so $E\delta_j^{(n)}(B) = P_n(B)$.

The counting measure is determined by the sums of δ-measures

$$\nu_n(B) := \sum_{j=1}^{n} \delta_j^{(n)}(B), \quad B \in \mathcal{X}, \tag{5.70}$$

therefore

$$\begin{aligned} E\nu_n(B) &= nP_n(B) \\ &=: \lambda_n(B). \end{aligned}$$

Investigate the U-statistics of sample (5.69) under the Poisson approximation scheme

$$\lim_{n\to\infty} nP_n(B) = \lambda(B), \quad B \in \mathcal{X}, \tag{5.71}$$

with some Radon measure $\lambda(B)$. In this case, the counting measure (5.70) is weakly convergent to some Poisson measure $\nu(B)$ with intensity $\lambda(B) = E\nu(B)$.

Theorem 5.5. *Assume that the following conditions hold:*

The degree of U-statistics increases as $n \to \infty$ in the following way:

$$b_n := m/n \longrightarrow b\,, \quad 0 < b < 1\,. \tag{5.73}$$

The degenerated kernels g_{mc} in (5.56) satisfy the following asymptotic property with some normalized coefficients $a_m \geq 0$

$$\lim_{n\to\infty} a_m g_{mc}(x_1, \dots, x_c) = g_c(x_1, \dots, x_c),$$

and the functions $g_c := g_c(x_1, \dots, x_c)$ are continuous with compact supports. The functions $a_m g_{mc}$ are also uniformly bounded

$$|a_m g_{mc}| \leq C < +\infty.$$

Then, in the Poisson approximation scheme (5.71), the weak convergence

$$a_m[U_{nm}^{\delta}(\Phi) - \widehat{\Phi}] \Rightarrow \sum_{c=r}^{\infty} \frac{b^c}{c!} I_c^{\nu}(g_c), \tag{5.74}$$

with Ito–Poisson stochastic integrals

$$I_c^{\nu}(g_c) := \int_{X^c} g_c(x_1, \dots, x_c) \prod_{s=1}^{c} \nu(dx_s),$$

takes place.

Proof. First, transform the Hoeffding decomposition of the U-statistics (5.56) to the form

$$a_m[U^\delta_{nm}(\Phi) - \widehat{\Phi}] = \sum_{c=r}^{m} \frac{m^{[c]}}{c!} U^\delta_{nc}(g_c) + \Delta_{n1} , \tag{5.75}$$

where

$$\Delta_{n1} := \sum_{c=r}^{m} \frac{m^{[c]}}{c!} U^\delta_{nc}(a_m g_{mc} - g_c).$$

Further, use the representation of U-statistics as a functional over a permanent random δ-measure (see (5.2) and (5.14)):

$$U^\delta_{nc}(g_c) = \int_{X^c} g_c(x_1, \dots, x_c) \pi^\delta_n(dx_1, \dots, dx_c)$$

with

$$\pi^\delta_n(dx_1, \dots, dx_c) := n^{-[c]} \text{Per } A^c_n(\delta(B)).$$

In addition, by theorem 2.3 (for k = m = c),

$$\begin{aligned} \text{Per } A^c_n(\delta(B)) &= \prod_{s=1}^{c} \nu_n(B_s) \\ &= \prod_{s=1}^{m} \sum_{j_s=1}^{(n-s+1)} \delta^{(n)}_{j_s}(B_s), \end{aligned}$$

where in each sum $j_s \neq j_1, \dots, j_{s-1}$, $2 \leq s \leq m$.

Thus the U-statistics $U^\delta_{nc}(g_c)$ have the representation

$$U^\delta_{nc}(g_c) = n^{-[c]} I^{\nu_n}_c(g_c) .$$

Furthermore, under the Poisson approximation scheme (5.71),

$$I^{\nu_n}_c(g_c) \Rightarrow I^\nu_c(g_c) , \quad n \to \infty.$$

Transform the first term in (5.75) as follows:

$$\begin{aligned} \sum_{c=r}^{m} \frac{m^{[c]}}{c!} U^\delta_{nc}(g_c) &= \sum_{c=r}^{m} \frac{m^{[c]}}{n^{[c]}} I^{\nu_n}_c(g_c)/c! \\ &= \sum_{c=r}^{m} b^c I^{\nu_n}_c(g_c)/c! + \Delta_{n2} , \end{aligned} \tag{5.76}$$

where

$$\Delta_{n2} := \sum_{c=r}^{m} \mathcal{E}_n(c) I^{\nu_n}_c(g_c)/c!,$$

$$\mathcal{E}(c) := m^{[c]}/n^{[c]} - b^c .$$

Consequently, under the conditions of theorem 5.5, the first term in (5.76) tends to the right-hand side of (5.74). In order to verify the statement of theorem 5.5 show that $\Delta_{nk}, k = 1, 2$ in (5.75) and (5.76) satisfy the asymptotic estimate

$$E|\Delta_{nk}| \longrightarrow 0 \quad \text{as} \quad m, n \to \infty \,, \quad k = 1, 2. \tag{5.77}$$

The first sum Δ_{n1} is transformed similarly to the first term in (5.75)

$$\Delta_{n1} = \sum_{c=r}^{m} \frac{m^{[c]}}{n^{[c]}} I_c^{\nu_n}(a_m g_{mc} - g_c)/c!.$$

Under condition (5.71), estimate

$$E|I_c^{\nu_n}(a_m g_{mc} - g_c)| \longrightarrow 0 \,.$$

It is now easy to verify that (5.77) holds true for Δ_{n1}. When considering Δ_{n2}, note that under condition (5.73) of theorem 5.5, as in subsection 3.1 (see (3.6)), the estimate

$$\mathcal{E}_n(c) \longrightarrow 0 \quad \text{as} \quad m, n \to \infty$$

is obtainable. Now obtain the estimate for Δ_{n2} as follows

$$\Delta_{n2} = \Delta'_{n2} + \Delta''_{n2} \,,$$

where

$$\Delta'_{n2} := \sum_{c=r}^{m_0} \mathcal{E}_n(c) I_c^{\nu_n}(g_c)/c! \longrightarrow 0 \quad \text{as} \quad n \to \infty$$

and

$$\Delta''_{n2} := \sum_{c=m_0+1}^{m} \mathcal{E}_n(c) I_c^{\nu_n}(g_c)/c! \longrightarrow 0 \quad \text{as} \quad m_0 \to \infty \,.$$

The last asymptotic estimates give (5.77), for the sum Δ_{n2}.

Corollary 5.3. *For the primitive kernel*

$$\Phi(x_1, \ldots, x_m) = \prod_{s=1}^{m} \varphi(x_s),$$

under the conditions of theorem 5.5 *and*

$$\lambda_n(\varphi) = n \int_X \varphi(x) P_n(dx) \to \int_X \varphi(x) \lambda(dx) = \lambda(\varphi) \neq 0,$$

the weak convergence as $n \to \infty$

$$\lambda_n^{-m}(\varphi) U_{nm}^{\delta}(\Phi) \Rightarrow \exp \int_X \ln(1 - bg(x)) \nu(dx) \tag{5.78}$$

takes place with $g(x) = 1 - \varphi(x)/\lambda(\varphi)$.

Proof. By the Hoeffding decomposition formula (5.56)

$$U^{\delta}_{nm}(\Phi) = \sum_{c=0}^{m} \frac{m^{[c]}}{c!} U^{\delta}_{nc}(g_{mc})$$

with

$$g_{mc} = [\lambda_n(\varphi)]^{m-c} \prod_{s=1}^{c} [\varphi(x_s) - \lambda_n(\varphi)].$$

Hence, under condition (5.71),

$$\lambda_n^{-m}(\varphi) g_{mc} \longrightarrow (-1)^c \prod_{s=1}^{c} g(x_s) = (-1)^c g_c, m \to \infty.$$

By theorem 5.5 the limiting result in (5.78) has the form (cf. 5.74)

$$\lambda_n^{-m}(\varphi) U^{\delta}_{nm}(\Phi) \Rightarrow \sum_{c=0}^{\infty} (-1)^c \frac{b^c}{c!} I^{\nu}_c(g_c) . \tag{5.79}$$

By the Ito–Poisson formula the right-hand side of (5.79) is equal to the right-hand side of (5.78).

Suppose that $E|\Phi(\xi_1^{(n)}, \ldots, \xi_m^{(n)})| < \infty$ and define functions $\Phi_n^v : X^m \to R$ by

$$\Phi_n^v(x_1, \ldots, x_m) = \int_X \cdots \int_X \Phi(y_1, \ldots, y_m) \prod_{s=1}^{m} (P_n(dy_s) + v\delta_s(dy_s)), \tag{5.80}$$

where v is an arbitrary fixed number, $|v| < \infty$. Introduce the family of functionals (with respect to the parameter v):

$$U^{\delta}_{nm}(\Phi_n^v) = n^{-[m]} \sum_{1 \le i_1 \ne \ldots \ne i_m \le n} \Phi_n^v(\xi_{i_1}^{(n)}, \ldots, \xi_{i_m}^{(n)}) . \tag{5.81}$$

So $U^{\delta}_{nm}(\Phi_n^v)$ is the integral functional from Φ respectively to the permanent measure $\pi_n^{P_n + v\delta}(B_1, \ldots, B_m)$. For $c = 0, 1, \ldots, m$ denote

$$\Phi_{mc}(x_1, \ldots, x_c) = \int_X \cdots \int_X \Phi(x_1, \ldots, x_c, x_{c+1}, \ldots, x_m) \prod_{s=c+1}^{m} P_n(dx_s)$$

and suppose that there exists the limit

$$\lim_{n \to \infty} \Phi_{mc}(x_1, \ldots, x_c) = \Phi_c(x_1, \ldots, x_c) . \tag{5.82}$$

Example 5.7. Let

$$\Phi(x_1, \ldots, x_m) = \Phi(x_1 + \ldots + x_m) ,$$

where $\Phi(x)$ is a continuous function with compact support. Then

$$\begin{aligned}\Phi_{mc}(x_1,\ldots,x_c) &= E\Phi(x_1+\ldots+x_c+\xi^{(n)}_{c+1}+\ldots+\xi^{(n)}_m) \\ &= E\Phi(x_1+\ldots+x_c+\int_X y\nu_{m-c}(dy)).\end{aligned}$$

Since, under conditions (5.71) and (5.73),

$$\nu_{m-c}(B)\Rightarrow\nu_{b\lambda}(B)\ ,\quad B\in\mathcal{X},$$

where $\nu_{b\lambda}(B)$ is the Poisson random measure with intensity $b\lambda(B)$ then, under the condition $n\to\infty$,

$$\Phi_{mc}(x_1,\ldots,x_c)\longrightarrow E\Phi(x_1+\ldots+x_c+\int_X y\nu_{b\lambda}(dy)).$$

Thus, in the given example,

$$\Phi_c(x_1,\ldots,x_c)=E\Phi(x_1+\ldots+x_c+\int_X y\nu_{b\lambda}(dy)).$$

Theorem 5.6. *Assume that conditions* (5.71) *and* (5.73) *are satisfied and*

$$|\Phi_c(x_1,\ldots,x_c)|\le\prod_{s=1}^{c}h(x_s)\ , \tag{5.83}$$

where $h(x)\ge 0$ is the continuous function with compact support. Also let, under $n\to\infty$,

$$\sum_{c=0}^{m}\frac{1}{c!}(v\frac{m}{n})^c\int_X\cdots\int_X|\Phi_{mc}(x_1,\ldots,x_c)-\Phi_c(x_1,\ldots,x_c)|\prod_{s=1}^{c}\lambda_n(dx_s)\to 0. \tag{5.84}$$

Then the weak convergence

$$U^\delta_{nm}(\Phi^v_n)\Rightarrow\sum_{c=0}^{\infty}\frac{(vb)^c}{c!}I^\nu_c(\Phi_c) \tag{5.85}$$

takes place.

Proof. For the beginning, according to (5.80), represent Φ^v_n as

$$\Phi^v_n(\xi^{(n)}_1,\ldots,\xi^{(n)}_m)=\sum_{c=0}^{m}\frac{v^c}{c!}\sum_{1\le i_1\ne\ldots\ne i_c\le m}\Phi_{mc}(\xi^{(n)}_{i_1},\ldots,\xi^{(n)}_{i_c}).$$

Substituting this formula in (5.81),

$$U^\delta_{nm}(\Phi^v_n)=\sum_{c=0}^{m}\frac{v^c}{c!}m^{[c]}n^{-[c]}\sum_{1\le i_1\ne\ldots\ne i_c\le m}\Phi_{mc}(\xi^{(n)}_{i_1},\ldots,\xi^{(n)}_{i_c}). \tag{5.86}$$

Write formula (5.86) as

$$U_{nm}^{v}(\Phi_n^{\delta}) = \sum_{c=0}^{m} \frac{v^c}{c!}\left(\frac{m}{n}\right)^c \sum_{1\le i_1\ne\cdots\ne i_c\le n} \Phi_c(\xi_{i_1}^{(n)},\ldots,\xi_{i_c}^{(n)})$$
$$+\Delta_{n1} + \Delta_{n2}, \tag{5.87}$$

where

$$\Delta_{n1} = \sum_{c=0}^{m} \frac{v^c}{c!}(m^{[c]}n^{-[c]} - m^c n^{-c})$$
$$\times \sum_{1\le i_1\ne\cdots\ne i_c\le n} \Phi_c(\xi_{i_1}^{(n)},\ldots,\xi_{i_c}^{(n)}),$$
$$\Delta_{n2} = \sum_{c=0}^{m} \frac{v^c}{c!} m^{[c]} n^{-[c]}$$
$$\times \sum_{1\le i_1\ne\cdots\ne i_c\le n} \left(\Phi_{mc}(\xi_{i_1}^{(n)},\ldots,\xi_{i_c}^{(n)}) - \Phi_c(\xi_{i_1}^{(n)},\ldots,\xi_{i_c}^{(n)})\right).$$

Taking into account (5.71)–(5.73) and condition (5.83) conclude that the limit of sums with respect to c coincides with the random value from the right-hand side of (5.85). That is why to prove (5.85) it is sufficient to check that, in (5.87), $\Delta_{ni} \to 0$ in probability. For this purpose write, taking into account (5.83),

$$E|\Delta_{n1}| \le \frac{1}{2m}\sum_{c=2}^{m} \frac{v^c}{(c-2)!}\left(\frac{m}{n}\right)^c$$
$$\times \int_X \cdots \int_X |\Phi_c(x_1,\ldots,x_c)| \prod_{s=1}^{c} \lambda_n(dx_s)$$
$$\le \frac{v^2}{2}\frac{m}{n^2}\sum_{c=0}^{m-2} \frac{v^c}{c!}\left(\frac{m}{n}\right)^c \lambda_n^c(h)$$
$$\le \frac{v^2}{2}\frac{m}{n^2}\exp\left(\frac{vm}{n}\lambda_n(h)\right) \to 0.$$

Further

$$E|\Delta_{n2}| \le \sum_{c=0}^{m} \frac{1}{c!}\left(v\cdot\frac{m}{n}\right)^c \int_X \cdots \int_X |\Phi_{mc}(x_1,\ldots,x_c)$$
$$-\Phi_c(x_1,\ldots,x_c)| \prod_{s=1}^{c} \lambda_n(dx_s) \to 0$$

by virtue of condition (5.84). As a result it proves statement (5.85).

5.5 Stable approximation scheme

The random measure $L_\alpha(\cdot)$ over the measurable space $(X, \mathcal{X})$ is called *a stable measure with index* $1 < \alpha < 2$, if random variable $L_\alpha(B)$, $B \in \mathcal{X}$ has a stable distribution with given index α and characteristic function

$$E\exp(itL_\alpha(B)) = \exp(-P(B)|t|^\alpha). \tag{5.88}$$

Introduce into consideration the random matrix

$$A_n^m(\eta) = [a_{ij} = \eta_j \delta_{\xi_j}(B_i);\ 1 \le i \le m,\ 1 \le j \le n], \tag{5.89}$$

where $B_i \in \mathcal{X}$, $\xi_1, \dots, \xi_n$ are independent random variables with equal distributions $P(\xi_j \in B) = P(B)$ and $\eta_1, \dots, \eta_n$ are independent random variables belonging to the domain of normal attraction of some α-stable law, $1 < \alpha < 2$; ξ_j and η_j are independent.

With respect to (5.89) define a random permanent measure

$$\pi_n^\alpha(B_1, \dots, B_m; \eta) = n^{-[m]/\alpha} \mathrm{Per}\ A_n^m(\eta) \tag{5.90}$$

and permanent integral functional

$$U_{mn}^\delta(\Phi, \eta) = \int_X \cdots \int_X \Phi(x_1, \dots, x_m) \pi_n^\alpha(dx_1, \dots, dx_m, \eta), \tag{5.91}$$

where $\Phi : X^m \to R$ is a symmetric function. If the integral in (5.91) is eliminated then

$$U_{mn}^\delta(\Phi, \eta) = n^{-[m]/\alpha} \sum_{1 \le i_1 \ne \dots \ne i_m \le n} \eta_{i_1} \cdots \eta_{i_m} \Phi(\xi_{i_1}, \dots, \xi_{i_m}). \tag{5.92}$$

The weak convergence (5.92) was studied in (Dehling et al. (1990)) where it was established that if $E|\Phi|^p < \infty$ with some $p > \alpha$ then the convergence

$$U_{mn}^\delta(\Phi, \eta) \Rightarrow I_m^{L_\alpha}(\Phi) \tag{5.93}$$

takes place, where

$$I_m^{L_\alpha}(\Phi) = \int_X \cdots \int_X \Phi(x_1, \dots, x_m) \prod_{s=1}^m L_\alpha(dx_s).$$

Demonstrate that, with the help of the "permanent" approach, the result of (5.93) could be deduced immediately from (5.91). Indeed, according to corollary 1.1,

$$\pi_n^\alpha(B_1, \dots, B_m; \eta) =$$

$$= n^{-1/\alpha} \sum_{q_1=1}^n \eta_{q_1} \delta_{\xi_{q_1}}(B_1) \dots (n-m+1)^{-1/\alpha} \sum_{q_m \in Q_1^{m-1}(q_1, \dots, q_{m-1})} \eta_{q_m} \delta_{\xi_{q_m}}(B_m).$$

Suppose that the sets $B_1, \ldots, B_m$ do not intersect each other: $B_i \cap B_j = \emptyset$, $i \neq j$. Then, taking into account this property (cf. (2.33)),

$$\pi_n^\alpha(B_1, \ldots, B_m, \eta) = n^{-[m]/\alpha} \prod_{s=1}^{m} (\sum_{j=1}^{n} \eta_j \delta_{\xi_j}(B_s)). \tag{5.94}$$

Here

$$n^{-1/\alpha} \sum_{j=1}^{n} \eta_j \delta_{\xi_j}(B_s) \Rightarrow L_\alpha(B_s) \tag{5.95}$$

because the characteristic function

$$E \exp\{itn^{-1/\alpha} \sum_{j=1}^{n} \eta_j \delta_{\xi_j}(B_s)\} =$$

$$= (1 + P(B_s))(E e^{itn^{-1/\alpha}\eta_1} - 1)^n \to \exp(-P(B_s)|t|^\alpha),$$

where $n \to \infty$. That is why, by virtue of (5.94)-(5.95),

$$\pi_n^\alpha(B_1, \ldots, B_m, \eta) \Rightarrow \prod_{s=1}^{m} L_\alpha(B_s). \tag{5.96}$$

Convergence (5.93) follows from (5.91), taking into account (5.96) and condition $E|\Phi|^p < \infty$, $p > \alpha$.

Remark 5.1. *The proof of* (5.93) *is possible to obtain also with the help of the permanent variant of composition method, where, instead of Gaussian measure W, substitute the stable measure L_α and, consequently, apply* (5.91).

Chapter 6

Mixed sampling permanents

Sampling matrices with differently distributed samples are considered under the Poisson and Gaussian approximation schemes.

6.1 Mixed sampling permanents with fixed degree

Let $\delta^{(n)} = (\delta_j^{(n)};\ 1 \leq j \leq n)$ be a sample satisfying the Poisson approximation condition

$$\begin{aligned} \mathcal{P}\{\delta_j^{(n)} = 1\} &= 1 - \mathcal{P}\{\delta_j^{(n)} = 0\} \\ &= \lambda/n, \quad \lambda > 0. \end{aligned} \tag{6.1}$$

Therefore the weak convergence

$$\nu_n = \sum_{j=1}^{n} \delta_j^{(n)} \Rightarrow \nu \quad \text{as} \quad n \to \infty \tag{6.2}$$

takes place, where ν is the Poisson distributed random variable with parameter $\lambda = E\nu$.

As was mentioned in section 2.1, the permanent of the sampling matrix

$$A_n^m(\delta^{(n)}) = [a_{ij} = \delta_j^{(n)};\ 1 \leq i \leq m, 1 \leq j \leq n]$$

has the value

$$\text{Per } A_n^m(\delta^{(n)}) = \nu_n^{[m]}. \tag{6.3}$$

Therefore, in the Poisson approximation scheme (6.1), the weak convergence

$$\text{Per } A_n^m(\delta^{(n)}) \Rightarrow \nu^{[m]} \quad \text{as} \quad n \to \infty \tag{6.4}$$

takes place. Considering the centered sample

$$\bar{\delta}^{(n)} = (\delta_j^{(n)} - \lambda/n,\ 1 \le j \le n),$$

according to theorem 2.1, the weak convergence

$$\text{Per } A_n^m(\bar{\delta}^{(n)}) \Rightarrow C_m(\nu, \lambda) \quad \text{as} \quad n \to \infty \tag{6.5}$$

takes place where

$$C_m(\nu, \lambda) = \sum_{r=0}^{m} (-\lambda)^{m-r} m^{[r]} \nu^{[r]} / r! \tag{6.6}$$

is the stochastic Charlier-Poisson polynomial.

Let $\varepsilon^{(n)} = (\varepsilon_j^{(n)},\ 1 \le j \le n)$ be a sample satisfying the Gaussian approximation condition

$$E\varepsilon_j^{(n)} = 0,$$

$$E(\varepsilon_j^{(n)})^2 = \sigma^2/n, \quad \sigma > 0. \tag{6.7}$$

According to the central limit theorem, the weak convergence

$$\sum_{j=1}^{n} \varepsilon_j^{(n)} \Rightarrow \sigma\tau \text{ as } \quad n \to \infty \tag{6.8}$$

takes place, where τ is the normal distributed standard random variable with $E\tau = 0$ and $E\tau^2 = 1$.

As is mentioned in section 4.2, the permanent of the sampling matrix

$$A_n^m(\varepsilon^{(n)}) = [a_{ij} = \varepsilon_j^{(n)};\ 1 \le i \le m, 1 \le j \le n]$$

converges weakly:

$$\text{Per } A_n^m(\varepsilon^{(n)}) \Rightarrow H_m(\sigma\tau) \quad \text{as} \quad n \to \infty \tag{6.9}$$

to the Hermite polynomial degree m.

Now investigate the asymptotic behaviour, as $n \to \infty$, of the permanents for the sum of two differently distributed samples.

Theorem 6.1. *Under the mixture condition* (6.1) *and* (6.7), *the weak convergences as* $n \to \infty$

$$\text{Per } A_n^m(\delta^{(n)} + \varepsilon^{(n)}) \Rightarrow \sum_{r=0}^{m} H_r(\sigma\tau) m^{[r]} \nu^{[m-r]} / r! \tag{6.10}$$

and

$$\text{Per } A_n^m(\bar{\delta}^{(n)} + \varepsilon^{(n)}) \Rightarrow \sum_{r=0}^{m} H_r(\sigma\tau) m^{[r]} C_{m-r}(\nu, \lambda) / r! \tag{6.11}$$

take place.

Proof. Using the decomposition formula (1.35)

$$\text{Per } A_n^m(\delta^{(n)}+\varepsilon^{(n)}) = \sum_{r=0}^{m} m^{[r]} \sum_{1\le q_1\ne\cdots\ne q_r\le n} \prod_{s=1}^{r} \varepsilon_{q_s}^{(n)} \text{Per } A_{n-r}^{m-r}(\delta^{(n)};q)/r! \qquad (6.12)$$

where matrix $A_{n-r}^{m-r}(\delta^{(n)};q)$ is constructed from matrix $A_n^{m-r}(\delta^{(n)})$ with deleted $q_1,\dots,q_r$ columns. Calculate

$$\text{Per } A_{n-r}^{m-r}(\delta^{(n)};q) = \nu_{n-r}^{[m-r]}(\bar{q}) \qquad (6.13)$$

where

$$\nu_{n-r}(\bar{q}) := \nu_n - \sum_{s=1}^{r} \delta_{q_s}^{(n)}.$$

Therefore getting the following representation

$$\text{Per } A_{n-r}^{m-r}(\delta^{(n)};q) = \nu_n^{[m-r]} - \Delta_n(q) \qquad (6.14)$$

where the difference

$$\Delta_n(q) := \nu_n^{[m-r]} - \nu_{n-r}^{[m-r]}(\bar{q}) \qquad (6.15)$$

satisfies the following estimate

$$E\Delta_n^2(q) \le C/n \to 0 \text{ as } \quad n\to\infty \qquad (6.16)$$

with the constant C, which can be chosen dependent only on λ and m.

Now formula (6.12) can be rewritten as follows

$$\text{Per } A_n^m(\delta^{(n)}+\varepsilon^{(n)}) = \sum_{r=0}^{m} m^{[r]} \sum_{1\le q_1\ne\cdots\ne q_r\le n} \prod_{s=1}^{r} \varepsilon_{q_s}^{(n)} \nu_n^{[m-r]}/r! + \Delta_n \qquad (6.17)$$

where it is easy to check that the random term Δ_n in (6.17) satisfies the estimate similar to (6.16)

$$E|\Delta_n|^2 \le C/n \to 0 \quad \text{as} \quad n\to\infty. \qquad (6.18)$$

The inner sum in (6.17) is the permanent of the sampling matrix of $r\times n$ dimension

$$\sum_{1\le q_1\ne\cdots\ne q_r\le n} \prod_{s=1}^{r} \varepsilon_{q_s}^{(n)} = \text{Per } A_n^r(\varepsilon^{(n)}).$$

Therefore the following representation takes place:

$$\text{Per } A_n^m(\delta^{(n)}+\varepsilon^{(n)}) = \sum_{r=0}^{m} m^{[r]} \text{Per } A_n^r(\varepsilon^{(n)}) \nu_n^{[m-r]}/r! + \Delta_n. \qquad (6.19)$$

Taking into account (6.4), (6.9) and (6.18) the weak limit in (6.19) gives result (6.10) of theorem 6.1.

To prove statement (6.11) employ decomposition formula (1.35) also to get the following representation

$$\text{Per } A_n^m(\bar{\delta}^{(n)} + \varepsilon^{(n)}) = \sum_{r=0}^{m} m^{[r]} \sum_{1 \le q_1 \ne \dots \ne q_r \le n} \prod_{s=1}^{r} \varepsilon_{q_s}^{(n)} \text{Per } A_{n-r}^{m-r}(\bar{\delta}^{(n)}; q)/r!. \tag{6.20}$$

Using decomposition formula (1.32) obtain

$$\text{Per } A_{n-r}^k(\bar{\delta}^{(n)}; q) = \sum_{c=0}^{k} g_n(c)(-\lambda)^{k-c} k^{[c]} \text{Per } A_{n-r}^c(\delta^{(n)}; q)/c! \tag{6.21}$$

where

$$g_n(c) := n^{[k]}/n^{[c]} n^{k-c} \to 1 \text{ as } \quad n \to \infty.$$

Using representation (6.14) rewrite (6.21)

$$\text{Per } A_{n-r}^k(\bar{\delta}^{(n)}; q) = \sum_{c=0}^{k} g_n(c)(-\lambda)^{k-c} k^{[c]} \nu_n^{[c]}/c! - \Delta_n \tag{6.22}$$

where Δ_n satisfies the estimate (6.18). Substituting (6.22) into (6.20) get

$$\text{Per } A_n^m(\bar{\delta}^{(n)} + \varepsilon^{(n)}) = \sum_{r=0}^{m} m^{[r]} \text{Per } A_n^r(\varepsilon^{(n)}) C_{m-r}^{(n)}(\nu, \lambda)/c! - \Delta_n \tag{6.23}$$

where

$$\begin{aligned} C_{m-r}^{(n)}(\nu; \lambda) \quad &:= \quad \sum_{c=0}^{m-r} g_n(c)(-\lambda)^{m-r-c}(m-r)^{[c]} \nu_n^{[c]}/c! \\ &\Rightarrow \quad C_{m-r}(\nu, \lambda) \end{aligned} \tag{6.24}$$

as $n \to \infty$. Taking the weak limit in (6.23) to get expression (6.11) from theorem 6.1.

The *mixed two-sampling matrix* is determined as follows

$$A_n^m(\varepsilon^{(n)}, \delta^{(n)}) = \begin{bmatrix} A_n^{m_1}(\varepsilon^{(n)}) \\ \\ A_n^{m_2}(\delta^{(n)}) \end{bmatrix},$$

$$m_1 + m_2 = m. \tag{6.25}$$

By the Laplace decomposition formula (1.11) the following representation of the permanent of matrices (6.25) is available,

$$\text{Per } A_n^m(\varepsilon^{(n)}, \delta^{(n)}) = m_1! \sum_{q \in Q_{m_1}} \prod_{s=1}^{m_1} \varepsilon_{q_s}^{(n)} \; m_2! \sum_{q' \in Q_{m_2}^{n-m_1}(q)} \prod_{s=1}^{m_2} \delta_{q'_s}^{(n)}. \tag{6.26}$$

Every sum of product in (6.26) is a symmetric polynomial .

Moreover, the second sum with factor $m_2!$ can be represented as the permanent of a sampling matrix dependent on the summation index $q = (q_1, \ldots, q_{m_1})$ in the following form

$$m_2! \sum_{q' \in Q_{m_2}^{n-m_1}(q)} \prod_{s=1}^{m_2} \delta_{q'_s}^{(n)} = \text{Per } A_{n-m_1}^{m_2}(\delta^{(n)}; q)$$

$$= \nu_{n-m_1}^{[m_2]}(q)$$

where, by the definition,

$$\nu_{n-m_1}(\bar{q}) := \nu_n - \nu_{m_1}(q)$$

$$\nu_{m_1}(q) := \sum_{s=1}^{m_1} \delta_{q_s}^{(n)}.$$

Considering the difference

$$\Delta_n(q, m_2) := \nu_n^{[m_2]} - \nu_{n-m_1}^{[m_2]}(\bar{q}) \tag{6.27}$$

and taking into account that $E\delta_j^{(n)} = \lambda/n$, it is easy to verify that the following estimate is valid

$$E\Delta_n^2(q, m_2) \leq c/n \to 0 \text{ as } \quad n \to \infty. \tag{6.28}$$

For this purpose first note that, for $|z| < 1$, the following decompositions are true

$$(1+z)^{\nu_n - \nu_{m_1}(q)} = \sum_{c \geq 0} z^c (\nu_n - \nu_{m_1}(q))^{[c]}/c!,$$

$$(1+z)^{\nu_n - \nu_{m_1}(q)} = (1+z)^{\nu_n}(1+z)^{-\nu_{m_1}(q)}$$

$$= \sum_{c \geq 0} z^c \sum_{d=0}^{c} (-\nu_{m_1}(q))^{[d]} \frac{1}{d!} \nu_n^{[c-d]} \frac{1}{(c-d)!}.$$

It is seen that

$$(\nu_n - \nu_{m_1}(q))^{[m_2]} = \sum_{d=0}^{m_2} (-\nu_{m_1}(q))^{[d]} \nu_n^{[m_2-d]} m_2^{[d]}/d!.$$

Isolate the term corresponding to $d = 0$ to find the formula

$$\Delta_n(q, m_2) = -\sum_{d=1}^{m_2} (-\nu_{m_1}(q))^{[d]} \nu_n^{[m_2-d]} m_2^{[d]}/d!.$$

Therefore it is easy to see that

$$E\Delta_n^2(q,m_2) \le m_2\sum_{d=1}^{m_2} E(-\nu_{m_1})^{2[d]}$$
$$\times E(\nu_n-\nu_{m_1})^{2[m_2-d]}m_2^{2[d]}/(d!)^2.$$

The following calculation proves when $\lambda \le n$

$$E(-\nu_{m_1})^{2[d]} \le \frac{\lambda}{n}m_1^2(-m_1)^{2[d]},$$

$$E(\nu_n-\nu_{m_1})^{2[m_2-d]} \le \sum_{c=0}^{m_2-d}(m_2-d)^{2[c]}/(c!)^2.$$

This reasoning proves inequality (6.28).

Note that the first sum in (6.26), with factor $m_1!$ separately, is the permanent of a sampling matrix

$$m_1!\sum_{q\in Q_{m_1}}\prod_{s=1}^{m_1}\varepsilon_{q_s}^{(n)} = \text{Per } A_n^{m_1}(\varepsilon^{(n)}).$$

Therefore obtain the following representation of the permanent of the mixed matrix

$$\text{Per } A_n^m(\varepsilon^{(n)},\delta^{(n)}) = \text{Per } A_n^{m_1}(\varepsilon^{(n)})\text{Per } A_n^{m_2}(\delta^{(n)}) + \Delta_n. \tag{6.29}$$

Lemma 6.1. *Under the mixed approximation conditions* (6.1) *and* (6.7), *the permanent of the mixed two-sampling matrix* (6.25) *is represented in form* (6.29) *with* Δ_n *satisfying the estimate*

$$E\Delta_n^2 \le c/n \to 0 \text{ as } n\to\infty. \tag{6.30}$$

Proof. The above mentioned analysis of formula (6.26) gives the following representation of the remainder term in (6.29):

$$\Delta_n = m_1!\sum_{q\in Q_{m_1}}\prod_{s=1}^{m_1}\varepsilon_{q_s}^{(n)}\,\Delta_n(q,m_2) \tag{6.31}$$

where $\Delta_n(q,m_2)$ satisfies estimate (6.28). Then, from (6.31), it is seen that

$$E\Delta_n^2 = m_1!n^{[m_1]}n^{-m_1}\sigma^{2m_1}$$

$$\times E\left(\sum_{d=1}^{m_2}(-\nu_{m_1})^{[d]}(\nu_n-\nu_{m_1})^{[m_2-d]}m_2^{[d]}/d!\right)^2.$$

Hence (6.30) follows.

The proof of lemma 6.1 is completed.

The asymptotic representation (6.28) of the permanent of a mixed two-sampling matrix gives an opportunity to get some new limiting results.

Theorem 6.2. *Under the mixed approximation conditions* (6.1) *and* (6.7) *the weak convergences, as* $n \to \infty$,

$$\text{Per } A_n^m(\varepsilon^{(n)}, \delta^{(n)}) \Rightarrow H_{m_1}(\sigma\tau)\nu^{[m_2]} \tag{6.32}$$

and

$$\text{Per } A_n^m(\varepsilon^{(n)}, \bar{\delta}^{(n)}) \Rightarrow H_{m_1}(\sigma\tau)C_{m_2}(\nu, \lambda) \tag{6.33}$$

take place.

Proof. The first statement (6.32) directly follows from representation (6.29).

To prove (6.33) it is necessary to verify that, similarly to (6.20), the following representation is true for the centered sample

$$\text{Per } A_n^m(\varepsilon^{(n)}, \bar{\delta}^{(n)}) = \text{Per } A_n^{m_1}(\varepsilon^{(n)})\text{Per } A_n^{m_2}(\bar{\delta}^{(n)}) + \bar{\Delta}_n \tag{6.34}$$

where the remainder term $\bar{\Delta}_n$, under the mixed conditions (6.1) and (6.7), satisfies the estimate

$$E\bar{\Delta}_n^2 \le c/n \to 0 \quad \text{as} \quad n \to \infty. \tag{6.35}$$

Then statement (6.33) will follow from the weak convergences (6.5) and (6.9).

Start from decomposition formula (1.11)

$$\text{Per } A_n^m(\varepsilon^{(n)}, \bar{\delta}^{(n)}) = m_1 \sum_{q \in Q_{m_1}} \prod_{s=1}^{m_1} \varepsilon_{q_s}^{(n)} m_2! \sum_{q' \in Q_{m_2}^{n-m_1}(q)} \prod_{s=1}^{m_2} \bar{\delta}_{q'_s}^{(n)} \tag{6.36}$$

where the second sum with factor $m_2!$ has the following representation

$$m_2! \sum_{q' \in Q_{m_2}^{n-m_1}(q)} \prod_{s=1}^{m_2} \bar{\delta}_{q_s}^{(n)} = \text{Per } A_{n-m_1}^{m_2}(\bar{\delta}^{(n)}; q). \tag{6.37}$$

Using decomposition formula (2.14)

$$\text{Per } A_{n-m_1}^{m_2}(\bar{\delta}^{(n)}; q) = \sum_{r=0}^{m_2} q_n(r)(-\lambda)^{m_2-r} m_2^{[r]} \nu_{n-m_1}^{[r]}(\bar{q})/r! \tag{6.38}$$

where

$$q_n(r) := (n - m_1 - r)^{[m_2-r]}/(n-m)^{m_2-r} \to 1 \quad \text{as} \quad \to \infty$$

for all $0 \le r \le m_2$.

Employ representation (6.27) and taking into account estimate (6.28) obtain the following representation

$$\text{Per } A^{m_2}_{n-m_1}(\bar{\delta}^{(n)}; q) = \text{Per } A^{m_2}_{n}(\bar{\delta}^{(n)}) + \Delta_n(q) \tag{6.39}$$

where the remainder term $\Delta_n(q)$ satisfies estimate (6.28). Substituting (6.39) into (6.36) verify, similarly to the proof of lemma 6.1, that the asymptotic representation (6.34) is valid.

6.2 Mixture approximation with increasing degrees

In this section the degree parameter m is supposed to be increasing to infinity together with $n \to \infty$.

Theorem 6.3. *Under the mixture approximation conditions* (6.1) *and* (6.7) *and with increasing degrees*

$$b_n := m/n \to b \text{ as } \quad n \to \infty,$$
$$0 < b < 1, \tag{6.40}$$

the weak convergences, as $n \to \infty$,

$$n^{-[m]}\text{Per } A^m_n(\varepsilon^{(n)} + \delta^{(n)} + 1) \Rightarrow \exp(b\tau - b^2/2)(1+b)^{\nu} \tag{6.41}$$

and

$$n^{-[m]}\text{Per } A^m_n(\varepsilon^{(n)} + \bar{\delta}^{(n)} + 1) \Rightarrow \exp(b\tau - b^2/2)e^{-b\lambda}(1+b)^{\nu} \tag{6.42}$$

take place.

Proof. First, similar to the proof of theorem 3.1, establish that the following asymptotic representation is valid:

$$n^{-[m]}\text{Per } A^m_n(\varepsilon^{(n)} + \delta^{(n)} + 1)$$

$$= \prod_{j=1}^{n}[1 + b(\varepsilon^{(n)}_j + \delta^{(n)}_j)] - \Delta_{n1} - \Delta_{n2} \tag{6.43}$$

where the remainder terms

$$\Delta_{n1} := \sum_{c=0}^{m} \varepsilon_n(c)\text{Per } A^c_n(\varepsilon^{(n)} + \delta^{(n)})/c!,$$

$$\Delta_{n2} := \sum_{c=m+1}^{n} b^c\text{Per } A^c_n(\varepsilon^{(n)} + \delta^{(n)})/c!,$$

$$\varepsilon_n(c) := b^c - m^{[c]}/n^{[c]}.$$

Similarly to the proof of theorem 3.1 verify that

$$|\varepsilon_n(c)| \leq c|b - b_n| + c^2/m \to 0 \quad \text{as} \quad m, n \to \infty.$$

Using decomposition representation (6.12), taking into account the conditions (6.1) and (6.7) and the independence of samples $\varepsilon^{(n)}$ and $\delta^{(n)}$, get the following estimate

$$E|\text{Per } A_n^c(\varepsilon^{(n)} + \delta^{(n)})| \leq \sum_{r=0}^{c} c^{[r]} \lambda^{c-r} E|\text{Per } A_n^r(\varepsilon^{(n)})|/r!.$$

By Hölder's inequality

$$\begin{aligned} [E|\text{Per } A_n^r(\varepsilon^{(n)})|]^2 &\leq E[\text{Per } A_n^r(\varepsilon^{(n)})]^2 \\ &\leq \sigma^{2r}. \end{aligned}$$

Therefore

$$\begin{aligned} E|\text{Per } A_n^c(\varepsilon^{(n)} + \delta^{(n)})| &\leq \sum_{r=0}^{c} c^{[r]} \lambda^{c-r} \sigma^r / r! \\ &\leq (\sigma + \lambda)^c. \end{aligned}$$

Now the remainder term Δ_{n1} allows the estimate

$$E|\Delta_{n1}| \leq (|b - b_n| + (\sigma + \lambda)/m)(\sigma + \lambda)e^{\sigma+\lambda} \to 0 \tag{6.44}$$

as $m, n \to \infty$. The estimation of the second remainder term Δ_{n_2} is more simple

$$\begin{aligned} E|\Delta_{n2}| &\leq \sum_{c=m+1}^{n} b^c E|\text{Per } A_n^c(\varepsilon^{(n)} + \delta^{(n)})|/c! \\ &\leq \sum_{c=m+1}^{n} b^c (\sigma + \lambda)^c / c! \\ &\leq \frac{[b(\sigma + \lambda)]^{m+1}}{(m+1)!} e^{b(\sigma+\lambda)} \to 0 \quad \text{as} \quad m, n \to \infty. \end{aligned} \tag{6.45}$$

Now establish that under the mixed conditions (6.1) and (6.7) the weak convergence

$$\prod_{j=1}^{n} [1 + b(\varepsilon^{(n)} + \delta^{(n)})] \Rightarrow \exp(b\tau - b^2/2)(1 + b)^{\nu} \text{ as } n \to \infty \tag{6.46}$$

takes place.

Introduce the truncated random variables

$$\hat{\varepsilon}_j^{(n)} := \varepsilon_j^{(n)} I(|\varepsilon_j^{(n)}| \leq c), \quad c < 1.$$

First establish the weak convergence (6.46) for the boundary random sample

$$\hat{\varepsilon}^{(n)} = (\hat{\varepsilon}_j^{(n)},\ 1 \leq j \leq n).$$

Considering the generating function of the sum of two independent samples in the following form

$$\prod_{j=1}^{n}[1 + b(\hat{\varepsilon}_j^{(n)} + \delta_j^{(n)})] = \prod_{j=1}^{n}(1 + b\hat{\varepsilon}_j^{(n)}) \prod_{j=1}^{n}(1 + b\delta_j^{(n)})\Delta_n(\hat{\varepsilon}^{(n)}, \delta^{(n)}) \tag{6.47}$$

where

$$\Delta_n(\hat{\varepsilon}^{(n)}, \delta^{(n)}) = \prod_{j=1}^{n}[1 - b^2\delta_j^{(n)}\hat{\varepsilon}_j^{(n)}/(1 + b\delta_j^{(n)})(1 + b\hat{\varepsilon}_j^{(n)})].$$

As a consequence to proposition 4.1 obtain

$$\prod_{j=1}^{n}(1 + b\hat{\varepsilon}_j^{(n)}) \Rightarrow \exp(b\tau - b^2/2) \quad \text{as} \quad n \to \infty. \tag{6.48}$$

It is obvious that the generating function of the sample $\delta^{(n)} = (\delta_j^{(n)},\ 1 \leq j \leq n)$ under the Poisson approximation condition (6.7) converges weakly

$$\begin{aligned}\prod_{j=1}^{n}(1 + b\delta_j^{(n)}) &= (1 + b)^{\nu_n} \\ &\Rightarrow (1 + b)^{\nu} \quad \text{as} \quad n \to \infty. \end{aligned} \tag{6.49}$$

To complete the proof of theorem 6.2 for boundary samples show that the third factor in (6.47) allows the following estimate

$$\Delta_n(\hat{\varepsilon}^{(n)}, \delta^{(n)}) \to 1 \quad \text{as} \quad n \to \infty. \tag{6.50}$$

Use an evident identity

$$1 - \prod_{j=1}^{n}(1 - \zeta_j) = \sum_{j=1}^{n} \zeta_j \prod_{s=j+1}^{n}(1 - \zeta_s)$$

with

$$\zeta_j := b^2\delta_j^{(n)}\hat{\varepsilon}_j^{(n)}/(1 + b\delta_j^{(n)})(1 + b\hat{\varepsilon}_j^{(n)})$$

which allows the estimate

$$E|\zeta_j| \leq a/n.$$

Therefore

$$\begin{aligned} E|\prod_{s=j+1}^{n}(1 - \zeta_s)| &\leq \prod_{s=j+1}^{n}(1 + E|\zeta_s|) \\ &\leq (1 + a/n)^n. \end{aligned}$$

Obtain the estimate

$$E|1-\Delta_n(\hat{\varepsilon}^{(n)},\delta^{(n)})| \leq \frac{a}{n}(1+\frac{a}{n})^n \to 0 \text{ as } n\to\infty.$$

The weak convergence (6.50) holds.

Analogously to the proof of lemma 4.1, it can be shown that

$$\begin{aligned}\Delta_n &:= \prod_{j=1}^{n}\left[1+b(\varepsilon_j^{(n)}+\delta_j^{(n)})\right]\\ &\quad -\prod_{j=1}^{n}\left[1+b(\hat{\varepsilon}_j^{(n)}+\delta_j^{(n)})\right] \xrightarrow{P} 0 \quad \text{as} \quad n\to\infty. \end{aligned} \tag{6.51}$$

First write the formula

$$\begin{aligned}E\Delta_n^2 &= E\prod_{j=1}^{n}(1+b(\varepsilon_j^{(n)}+\delta_j^{(n)}))^2\\ &\quad -E\prod_{j=1}^{n}(1+b(\varepsilon_j^{(n)}+\delta_j^{(n)}))(1+b(\hat{\varepsilon}_j^{(n)}+\delta_j^{(n)}))\\ &\quad +E\prod_{j=1}^{n}(1+b(\hat{\varepsilon}_j^{(n)}+\delta_j^{(n)}))^2.\end{aligned}$$

By virtue of the independence of $\xi_j^{(n)}$ and $\delta_j^{(n)}$ obtain

$$\begin{aligned}E\Delta_n^2 &= (E(1+b(\varepsilon_1^{(n)}+\delta_1^{(n)}))^2)^n\\ &\quad -2\Big(E(1+b(\varepsilon_1^{(n)}+\delta_1^{(n)}))(1+b(\hat{\varepsilon}_1^{(n)}+\delta_1^{(n)}))\Big)^n\\ &\quad +\Big(E(1+b(\hat{\varepsilon}_1^{(n)}+\hat{\delta}_1^{(n)}))^2\Big)^n.\end{aligned}$$

Hence it is seen that $E\Delta_n^2 \to 0$ when $n\to\infty$, i.e. (6.51) is true.

The weak convergence (6.51) completes the proof of theorem 6.2. To prove statement (6.42) of theorem 6.3, consider the representation of the permanents with shift with the help of decomposition formula (1.22)

$$\begin{aligned}n^{-[m]}\text{Per } A_n^m(\varepsilon^{(n)} &+ \bar{\delta}^{(n)}+1)\\ &= n^{-[m]}\text{Per } A_n^m(\varepsilon^{(n)}+\delta^{(n)}+1-\lambda/n)\\ &= \sum_{c=0}^{m}(1-\lambda/n)^{m-c}m^{[c]}n^{-[c]}\text{Per } A_n^c(\varepsilon^{(n)}+\delta^{(n)})/c!\\ &= (1-\lambda/n)^m n^{-[m]}\text{Per } A_n^m(\varepsilon^{(n)}+\delta^{(n)}+1)+\Delta_n \end{aligned} \tag{6.52}$$

where

$$\Delta_n := \sum_{c=1}^{m} q_n(c) m^{[c]} n^{-[c]} \text{Per } A_n^c(\varepsilon^{(n)} + \delta^{(n)})/c!$$

with

$$\begin{aligned} q_n(c) &:= (1-\lambda/n)^{m-c} - (1-\lambda/n)^m \\ &\leq \lambda c/n. \end{aligned}$$

Therefore obtain the following estimate

$$E|\Delta_n| \leq \lambda(\sigma+\lambda)e^{\sigma+\lambda}/n \to 0 \text{ as } n \to \infty. \tag{6.53}$$

Note that under condition (6.40) of theorem 6.3

$$(1-\lambda/n)^m \to e^{-b\lambda} \quad \text{as } n \to \infty. \tag{6.54}$$

Representation (6.52), together with the estimates (6.53) and (6.54), completes the proof (6.42) of theorem 6.3.

Now consider the permanent of a two-sampling matrix with increasing degree of one sample.

Theorem 6.4. *Under the mixture conditions* (6.1) *and* (6.7) *and with increasing degree*

$$m_2/n \to b \quad \text{as} \quad n \to \infty, \quad 0 < b < 1, \tag{6.55}$$

the weak convergences, as $n \to \infty$,

$$n^{-[m_2]} \text{ Per } A_n^m(\varepsilon^{(n)}, \delta^{(n)} + 1) \Rightarrow H_{m_1}(\sigma\tau)(1+b)^\nu \tag{6.56}$$

and

$$n^{-[m_2]} \text{ Per } A_n^m(\varepsilon^{(n)}, \bar{\delta}^{(n)} + 1) \Rightarrow H_{m_1}(\sigma\tau)e^{-b\lambda}(1+b)^\nu \tag{6.57}$$

take place.

Proof. Decomposition formula (1.44) gives the following representation

$$n^{-[m_2]} \text{ Per } A_n^m(\varepsilon^{(n)}, \delta^{(n)} + 1) = \sum_{c=0}^{m_2} m_2^{[c]} n^{-[c]} \text{Per } A_n^{m_1+c}(\varepsilon^{(n)}, \delta^{(n)})/c! \tag{6.58}$$

With the help of lemma 6.1 get the following asymptotic representation

$$\begin{aligned} n^{-[m_2]}\text{Per } A_n^m(\varepsilon^{(n)}, \delta^{(n)} + 1) &= \sum_{c=0}^{m_2} m_2^{[c]} n^{-[c]} \\ &\times \text{Per } A_n^c(\delta^{(n)}) \text{Per } A_n^{m_1}(\varepsilon^{(n)})/c! + \Delta_n \end{aligned} \tag{6.59}$$

where $\Delta_n \xrightarrow{P} 0$ as $n \to \infty$. The sum in (6.59) is transformed similar to the proof of theorem 6.2 as follows

$$n^{-[m_2]}\text{Per } A_n^m(\varepsilon^{(n)}, \delta^{(n)} + 1) = \text{Per } A_n^{m_1}(\varepsilon^{(n)}) \prod_{j=1}^{n}(1 + b\delta_j^{(n)}) + \Delta_n \tag{6.60}$$

where $\Delta_n \xrightarrow{P} 0$ as $n \to \infty$.

Taking into account the weak limit (6.60), consider the weak convergences (6.9) and (6.49) and obtain the statement (6.56) of theorem 6.4.

The proof of (6.57) is analogous to the proof of theorem 6.4., but instead of formula (6.58) use the following representation

$$n^{-[m_2]}\text{Per } A_n^m(\varepsilon^{(n)}, \bar{\delta}^{(n)} + 1)$$
$$= \sum_{c=0}^{m_2}(1 - \lambda/n)^{m_2-c} m_2^{[c]} n^{-[c]}\text{Per } A_n^{m_1+c}(\varepsilon^{(n)}, \delta^{(n)})/c!. \tag{6.61}$$

Another type of limiting result is when the degree of the normal approximating sample is increasing.

Theorem 6.5. *Under the Poisson approximation condition* (6.7) *and the normal approximation condition of the following form*

$$E\varepsilon_j^{(n)} = \mu/\sqrt{n},$$

$$\mu \neq 0,$$

$$E[\varepsilon_j^{(n)} - \mu/\sqrt{n}]^2 = \sigma^2/n \tag{6.62}$$

and with increasing degree

$$m_1/\sqrt{n} \to \gamma \quad as \quad n \to \infty, \quad \gamma > 0, \tag{6.63}$$

the weak convergence, as $n \to \infty$,

$$n^{-[m_1]}(\mu/\sqrt{n})^{-m_1}\text{Per } A_n^m(\varepsilon^{(n)}, \delta^{(n)}) \Rightarrow \exp[a\tau - a^2/2]\nu^{[m_2]}, \tag{6.64}$$

where $a = \gamma\sigma/\mu$, *takes place.*

Proof. Introduce the centered sample

$$\xi^{(n)} = (\xi_j^{(n)} = \varepsilon_j^{(n)} - \mu/\sqrt{n}, \ 1 \leq j \leq n)$$

with $E\xi_j^{(n)} = 0$ and $E(\xi_j^{(n)})^2 = \sigma^2/n$.

Therefore by the central limit theorem

$$\sum_{j=1}^{n}\xi_j^{(n)} \Rightarrow \sigma\tau \quad \text{as} \quad n \to \infty. \tag{6.65}$$

Decomposition formula (1.43) gives the following representation

$$n^{-[m_1]}(\mu/\sqrt{n})^{-m_1}\text{Per } A_n^m(\varepsilon^{(n)},\delta^{(n)})$$

$$= \sum_{c=0}^{m_1} q_n(c)\mu^{-c} m_1^{[c]} n^{-c/2}\text{Per } A_n^m(\xi^{(n)},\delta^{(n)})/c! \tag{6.66}$$

where

$$q_n(c) := n^c/n^{[c]} \to 1 \quad \text{as} \quad \to \infty.$$

Using lemma 6.1

$$n^{-[m_1]}(\mu/\sqrt{n})^{-m_1}\text{Per } A_n^m(\varepsilon^{(n)},\delta^{(n)})$$

$$= \sum_{c=0}^{m_1} q_n(c)\mu^{-c} m_1^{[c]} n^{-c/2}\text{Per } A_n^c(\xi^{(n)})\text{Per } A_n^{m_2}(\delta^{(n)})/c! + \Delta_n, \tag{6.67}$$

where Δ_n satisfies estimate (6.60).
The sum in (6.67) can be transformed similar to the analogous one in the proof of theorem 6.2 as follows,

$$\sum_{c=0}^{m_1} q_n(c)\mu^{-c} m_1^{[c]} n^{-c/2}\text{Per } A_n^c(\xi^{(n)})/c!$$

$$= \prod_{j=1}^{n}(1+\mu^{-1}\gamma\xi_j^{(n)}) + \Delta_n. \tag{6.68}$$

The asymptotic representations (6.67) and (6.68) give

$$n^{-[m_1]}(\mu/\sqrt{n})^{-m_1}\text{Per } A_n^m(\varepsilon^{(n)},\delta^{(n)})$$

$$= \prod_{j=1}^{n}(1+\mu^{-1}\gamma\xi_j^{(n)})\text{Per } A_n^{m_2}(\delta^{(n)}) + \Delta_n. \tag{6.69}$$

Taking into account (6.65) obtain

$$\prod_{j=1}^{n}(1+\mu^{-1}\gamma\xi_j^{(n)}) \Rightarrow \exp[a\tau - a^2/2] \quad \text{as} \quad n\to\infty$$

and, as usual, under condition (6.7)

$$\begin{aligned}\text{Per } A_n^{m_2}(\delta^{(n)}) &= \nu_n^{[m_2]}\\ &\Rightarrow \nu^{[m_2]} \quad \text{as} \quad n\to\infty.\end{aligned}$$

Taking the weak limit in (6.69) complete the proof of theorem 6.5.

Now consider the permanent of a two-sampling matrix with increasing degrees of both samples.

Theorem 6.6. *Let the Poisson approximation condition (6.7) and the normal approximation be valid.*

Let the degrees of samples m_1 and m_2 satisfy the following asymptotic relations

$$m_1/\sqrt{n} \to \gamma,$$

$$m_2/n \to b \text{ as } n \to \infty,$$

$$\gamma > 0, \qquad 0 < b < 1. \tag{6.70}$$

Then the weak convergences, as $n \to \infty$,

$$n^{-[m]}(\mu/\sqrt{n})^{-m_1} \operatorname{Per} A_n^m(\varepsilon^{(n)}, \delta^{(n)} + 1)$$

$$\Rightarrow \exp[a\tau - a^2/2](1+b)^{\nu}, \tag{6.71}$$

$$n^{-[m]}(\mu/\sqrt{n})^{-m_1} \operatorname{Per} A_n^m(\varepsilon^{(n)}, \delta^{(n)} + 1) \Rightarrow \exp[a\tau - a^2/2]e^{-b\lambda}(1+b)^{\nu} \tag{6.72}$$

take place with $a = \gamma\sigma/\mu$.

Proof. The limiting results of theorem 6.6 are the consequence of the following asymptotic representations

$$n^{-[m]}(\mu/\sqrt{n})^{-m_1} \operatorname{Per} A_n^m(\varepsilon^{(n)}, \delta^{(n)} + 1) = \prod_{j=1}^{n}(1 + \mu^{-1}\gamma\xi_j^{(n)} + b\delta_j^{(n)}) + \Delta_n \tag{6.73}$$

and

$$\operatorname{Per} A_n^m(\varepsilon^{(n)}, \bar{\delta}^{(n)} + 1) = (1 - \lambda/n)^{m_2} \operatorname{Per} A_n^m(\varepsilon^{(n)}, \delta^{(n)} + 1)$$

$$+ n^{[m]}(\mu/\sqrt{n})^{m_1}\Delta_n \tag{6.74}$$

where the residual terms Δ_n satisfy the estimate

$$\Delta_n \xrightarrow{P} 0 \quad \text{as} \quad n \to \infty. \tag{6.75}$$

At first establish that relation (6.74) is true. By the decomposition formula (1.34), for the case of $k = 2$, obtain

$$n^{-[m]} \operatorname{Per} A_n^m(\varepsilon^{(n)}, \bar{\delta}^{(n)} + 1)$$

$$= n^{-[m]}\mathrm{Per}A_n^m(\xi^{(n)} + \mu/\sqrt{n}, \delta^{(n)} + 1 - \lambda/n)$$

$$= \sum_{c=0}^{m} n^{-[c]} \sum_{r_1+r_2=c} m_1^{[r_1]} m_2^{[r_2]} (\mu/\sqrt{n})^{m_1-r_1}(1-\lambda/n)^{m_2-r_2}$$

$$\times \mathrm{Per}A_n^c(\xi^{(n)}, \delta^{(n)})/r_1!r_2!$$

$$= (1-\lambda/n)^{m_2} \sum_{c=0}^{m} n^{-[c]} \sum_{r_1+r_2=c} m_1^{[r_1]} m_2^{[r_2]} (\mu/\sqrt{n})^{m_1-r_1}$$

$$\times \mathrm{Per}A_n^c(\xi^{(n)}, \delta^{(n)})/r_1!r_2! + \Delta_n$$

$$= (1-\lambda/n)^{m_2} n^{-[m]}\mathrm{Per}A_n^m(\varepsilon^{(n)}, \delta^{(n)} + 1) + \Delta_n \ ,$$

where

$$\Delta_n := \sum_{c=1}^{m} n^{-[c]} \sum_{r_1+r_2=c} \alpha_n(r_2) m_1^{[r_1]} m_2^{[r_2]} (\mu/\sqrt{n})^{m_1-r_1}$$

$$\times \mathrm{Per}A_n^c(\xi^{(n)}, \delta^{(n)})/r_1!r_2!,$$

$$\alpha_n(r_2) \ := \ (1-\lambda/n)^{m_2-r_2} - (1-\lambda/n)^{m_2}$$

$$\leq \ r_2\lambda/n.$$

By the Hölder inequality, for $c = r_1 + r_2$,

$$[E|\mathrm{Per}A_n^c(\xi^{(n)}, \delta^{(n)})|]^2 \leq E[\mathrm{Per}A_n^{r_1}(\xi^{(n)})]^2 E[\mathrm{Per}A_n^{r_2}(\delta^{(n)})]^2,$$

where

$$E[\mathrm{Per}A_n^{r_1}(\xi^{(n)})]^2 \leq (\sigma/\mu)^{2r_1},$$

$$E[\mathrm{Per}A_n^{r_2}(\delta^{(n)})]^2 \leq (1+\lambda)^{2r_2}.$$

It implies the estimate

$$E|\Delta_n| \leq C/n \to 0 \quad \text{as} \quad n \to \infty \tag{6.76}$$

Now, using decomposition formula (1.34) in the same way:

$$n^{-[m]}(\mu/\sqrt{n})^{-m_1}\mathrm{Per}\ A_n^m(\varepsilon^{(n)}, \delta^{(n)} + 1)$$

$$= n^{-[m]}(\mu/\sqrt{n})^{-m_1}\mathrm{Per}\ A_n^m(\xi^{(n)} + \mu/\sqrt{n}, \delta^{(n)} + 1)$$

$$= \sum_{c=0}^{m} \sum_{r_1+r_2=c} q_n(r_1, r_2) \mathrm{Per} A_n^c(\xi^{(n)}/\mu, \delta^{(n)})/r_1!r_2! \ , \tag{6.77}$$

where

$$q_n(r_1, r_2) := n^{r_1/2} m_1^{[r_1]} m_2^{[r_2]} / n^{[c]}. \tag{6.78}$$

It is easy to check that, under the conditions of theorem 6.6,

$$\alpha_n(r_1, r_2) := q_n(r_1, r_2) - \gamma^{r_1} b^{r_2} \to 0 \quad \text{as} \quad n \to \infty. \tag{6.79}$$

So the normalized permanent in theorem 6.6 can be represented as follows

$$n^{-[m]} (\mu/\sqrt{n})^{-m_1} \mathrm{Per} A_n^m(\varepsilon^{(n)}, \delta^{(n)} + 1)$$

$$= \sum_{c=0}^{n} \sum_{r_1+r_2=c} (\gamma/\mu)^{r_1} b^{r_2} \mathrm{Per} A_n^c(\xi^{(n)}, \delta^{(n)})/r_1!r_2! + \Delta_{n1} - \Delta_{n2} \ , \tag{6.80}$$

where

$$\Delta_{n1} := \sum_{c=1}^{m} \sum_{r_1+r_2=c} \alpha_n(r_1, r_2) \mathrm{Per} A_n^c(\xi^{(n)}, \delta^{(n)})/r_1!r_2! \ , \tag{6.81}$$

and

$$\Delta_{n2} := \sum_{c=m+1}^{n} \sum_{r_1+r_2=c} (\gamma/\mu)^{r_1} b^{r_2} \mathrm{Per} A_n^c(\xi^{(n)}, \delta^{(n)})/r_1!r_2! \ . \tag{6.82}$$

From which the next estimate follows

$$E|\Delta_{nk}| \to 0 \quad \text{as} \quad n \to \infty, \quad k = 1, 2 \ . \tag{6.83}$$

Note that the first sum in (6.80) is the generating function of the permanents of the two-sampling matrices

$$\sum_{c=0}^{n} \sum_{r_1+r_2=c} (\gamma/\mu)^{r_1} b^{r_2} \mathrm{Per} A_n^c(\xi^{(n)}, \delta^{(n)})/r_1!r_2!$$

$$= \prod_{j=1}^{n} (1 + (\gamma/\mu)\xi_j^{(n)} + b\delta_j^{(n)}) \ . \tag{6.84}$$

Representations (6.80) – (6.84) yield the weak convergence (6.71) of theorem 6.6. The weak convergence (6.72) is the consequence of relations (6.71), (6.74) and (6.75) due to the convergence

$$(1 - \lambda/n)^{m_2} \to e^{-b\lambda} \quad \text{as} \quad n \to \infty.$$

6.3 Mixed multiple sampling matrix

The object under consideration in this section is the sum of two multiple sampling matrices

$$A_n^m(\varphi^{(n)} + \varepsilon^{(n)}) = A_n^m(\varphi_1^{(n)}, \dots, \varphi_k^{(n)}) + A_n^m(\varepsilon_1^{(n)}, \dots, \varepsilon_k^{(n)})$$

where the multiple sampling matrix $A_n^m(\varphi_1^{(n)}, \dots, \varphi_k^{(n)})$ contains k blocks of sampling matrices

$$A_n^{m_s}(\varphi_s^{(n)}) = [a_{ij} = \varphi_s(\xi_j^{(n)});\ 1 \le i \le m_s,\ 1 \le j \le n],\ 1 \le s \le k,$$

constructed with given continuous finite real valued functions $\varphi_s : X \to R$ and with the sample $\xi^{(n)} = (\xi_j^{(n)}, 1 \le j \le n)$ of random variables taking values on a measurable space $(X, \mathcal{X})$ and satisfying the Poisson approximation condition

$$\mathcal{P}\{\xi_j^{(n)} \in B\} = \lambda(B)/n, \quad B \in \mathcal{X} \tag{6.85}$$

with some Radon finite measure $\lambda(B)$.

The multiple sampling matrix $A_n^m(\varepsilon_1^{(n)}, \dots, \varepsilon_k^{(n)})$ contains k blocks of sampling matrices

$$A_n^{m_s}(\varepsilon_s^{(n)}) = [a_{ij} = \varepsilon_{sj}^{(n)};\ 1 \le i \le m_s, 1 \le j \le k]$$

constructed with mutually independent samples

$$\varepsilon_s^{(n)} = (\varepsilon_{s_j}^{(n)},\ 1 \le j \le n), \quad 1 \le s \le k,$$

satisfying the normal approximation condition

$$E\varepsilon_{sj}^{(n)} = 0,$$

$$E(\varepsilon_{sj}^{(n)})^2 = \sigma_s^2/n,$$

$$\sigma_s > 0. \tag{6.86}$$

As usual, $m = m_1 + \dots + m_k$.

Theorem 6.7. *Under the mixed approximation conditions* (6.85) *and* (6.86) *and for fixed degrees* m_s, $1 \le s \le k$, *the weak convergences, as* $n \to \infty$,

$$\text{Per } A_n^m(\varphi^{(n)} + \varepsilon^{(n)}) \Rightarrow \sum_{c=0}^{m} \sum_{r_1 + \dots + r_k = c} I_c^\nu(\varphi_1, \dots, \varphi_k) \times \prod_{s=1}^{k} H_{m_s - r_s}(\sigma_s \tau_s) m_s^{[r_s]}/r_s! \tag{6.87}$$

and

$$\text{Per } A_n^m(\bar\varphi^{(n)} + \varepsilon^{(n)}) \Rightarrow \sum_{c=0}^{m} \sum_{r_1+\cdots+r_k=c} I_c^{\bar\nu}(\varphi_1, \ldots, \varphi_k) \times \prod_{s=1}^{k} H_{m_s-r_s}(\sigma_s\tau_s) m_s^{[r_s]}/r_s! \tag{6.88}$$

take place. Here

$$\bar\varphi_s^{(n)} = \varphi_s(\xi_j^{(n)}) - \lambda(\varphi_s)/n,$$

$$\lambda(\varphi_s) := \int_X \varphi_s(x)\lambda(dx).$$

Proof. Employing decomposition formula (1.40) with $a_s = \varphi_s^{(n)}$ and $b_s = \varepsilon_s^{(n)}$ it can be verified that the last term in (1.40) has the following limit representation by theorem 4.5.

$$\text{Per } A_{n-c}^{m-c}(\varepsilon_1^{(n)}, \ldots, \varepsilon_k^{(n)}; q) \Rightarrow \prod_{s=1}^{k} H_{m_s-r_s}(\sigma_s\tau_s), \tag{6.89}$$

as $n \to \infty$. The inner sums in (1.40), without the last term, have the following representation by theorem 1.1.

$$\prod_{s=1}^{k} m_s^{[r_s]} \sum_{q^{(1)}\in Q_{r_1}} \prod_{l=1}^{r_1} \varphi_1(\xi_{q_l^{(1)}}^{(n)}) \;\cdots \sum_{q^{(k)}\in Q_{r_k}^{n-r_k}(q^{(1)},\ldots,q^{(k-1)})} \prod_{l=1}^{r_k} \varphi_k(\xi_{q_l^{(k)}}^{(n)})$$

$$= \text{Per } A_n^c(\varphi_1^{(n)}, \ldots, \varphi_k^{(n)}).$$

By theorem 2.4

$$\text{Per } A_n^c(\varphi_1^{(n)}, \ldots, \varphi_k^{(n)}) \Rightarrow I_c^{\nu}(\varphi_1, \ldots, \varphi_k). \tag{6.90}$$

Representation (6.89) and weak convergence (6.90) by formula (1.40) gives statement (6.87) of theorem 6.7.

The proof of (6.88) is analogous to the proof of (6.87) taking into account that the inner sums in (1.40), without the last term, are represented as follows

$$\text{Per } A_n^c(\bar\varphi_1^{(n)}, \ldots, \bar\varphi_k^{(n)}) \Rightarrow I_c^{\bar\nu}(\varphi_1, \ldots, \varphi_k)$$

by theorem 2.4.

Theorem 6.8. *Under the mixed approximation conditions* (6.85) *and* (6.86) *and with increasing degrees*

$$m_s/n \to b_s,$$

$$0 < b_s < 1,$$

$$1 \leq s \leq k,$$

the weak convergence, as $n \to \infty$,

$$n^{-[m]}\text{Per } A_n^m(1 + \varphi^{(n)} + \varepsilon^{(n)}) \Rightarrow$$

$$\Rightarrow \exp[\sum_{s=1}^{k}(b_s\tau_s - b_s^2/2) + \int_X \ln[1 + \sum_{s=1}^{k} b_s\varphi_s(x)]\nu(dx)] \tag{6.91}$$

takes place.

Proof. By decomposition formula (1.33)

$$n^{-[m]}\text{Per } A_n^m(1 + \varepsilon^{(n)} + \varphi^{(n)}) =$$

$$= \sum_{c=0}^{m} n^{-[c]} \sum_{r_1+\cdots+r_k=c} \prod_{s=1}^{k} m_s^{[r_s]} \text{ Per } A_n^c(\varepsilon^{(n)} + \varphi^{(n)})/r_s!.$$

After this transformation the following representation is produced

$$n^{-[m]}\text{Per } A_n^m(1 + \varepsilon^{(n)} + \varphi^{(n)}) =$$

$$= \prod_{j=1}^{n}[1 + \sum_{s=1}^{k} b_s(\varepsilon_{s_j}^{(n)} + \varphi_s(\xi_j^{(n)}))] + \Delta_n \tag{6.92}$$

where $\Delta_n \xrightarrow{P} 0$ as $n \to \infty$.

Under approximation conditions (6.85) and (6.86) the generating sampling function in (6.92) has the weak limit, as $n \to \infty$, in the form of the right-hand side of (6.91).

Chapter 7

Rate of convergence

The rates of weak convergence of permanents in Poisson and Gaussian approximation schemes are obtained.

7.1 Rate of Poisson approximation

In the same way as in chapter 2 start from a sample $\delta_j^{(n)}, 1 \le j \le n$, in the series scheme with

$$\begin{aligned} \mathcal{P}\{\delta_j^{(n)} = 1\} &= 1 - \mathcal{P}\{\delta_j^{(n)} = 0\} \\ &= \lambda/n, \quad \lambda > 0. \end{aligned} \tag{7.1}$$

Let

$$\nu_n := \sum_{j=1}^{n} \delta_j^{(n)}$$

and ν be a Poisson distributed random variable with parameter λ:

$$\mathcal{P}\{\nu = k\} = e^{-\lambda}\lambda^k/k!, \ k \ge 0.$$

As was mentioned in section 2

$$\begin{aligned} \nu_n^{[m]} &:= \nu_n(\nu_n - 1)\dots(\nu_n - m + 1) \\ &= \text{Per } A_n^m(\delta^{(n)}), \end{aligned}$$

where

$$A_n^m(\delta^{(n)}) := [a_{ij} = \delta_j^{(n)}; \ 1 \le i \le m, 1 \le j \le n]$$

is a sampling matrix constructed on the sample

$$\delta^{(n)} := (\delta_j^{(n)}, \ 1 \le j \le n).$$

It is a well-known fact (see Barbour et al. (1992)) that, in the Poisson approximation scheme (7.1) under fixed m and for all k, $0 \leq k \leq n^{[m]}$,

$$|\mathcal{P}\{\nu_n^{[m]} = k\} - \mathcal{P}\{\nu^{[m]} = k\}| \to 0 \qquad (7.2)$$

as $n \to \infty$. Consider the problem of the estimate of the rate of decrease to zero in (7.2), relative to n.

Theorem 7.1. *Under the Poisson approximation condition* (7.1), *the following estimate is valid for all* $1 \leq m \leq n$ *and* $0 < \lambda \leq n$*:*

$$\left|\mathcal{P}\{\nu_n^{[m]} = k\} - \mathcal{P}\{\nu^{[m]} = k\}\right| \leq (1 - e^{-\lambda})\lambda/n. \qquad (7.3)$$

Proof. Let, as usual, I (A) denote the indicator function of the set $A \in \mathcal{X}$. Then

$$\begin{aligned} \mathcal{P}(\nu_n^{[m]} = k) &= EI(\nu_n^{[m]} = k) \\ &= \sum_{r=0}^{\infty} I(r^{[m]} = k)\mathcal{P}(\nu_n = r), \end{aligned}$$

$$\begin{aligned} \mathcal{P}(\nu^{[m]} - k) &- EI(\nu^{[m]} - k) \\ &= \sum_{r=0}^{\infty} I(r^{[m]} = k)\mathcal{P}(\nu = r). \end{aligned}$$

So as to get the following estimate

$$|\mathcal{P}(\nu_n^{[m]} = k) - (\nu^{[m]} = k)| \leq$$

$$\leq \sum_{r=0}^{\infty} I(r^{[m]} = k)|\mathcal{P}(\nu_n = r) - \mathcal{P}(\nu = r)|. \qquad (7.4)$$

In the Poisson approximation scheme (7.1), the rate of convergence has the following form (Barbour et al. (1984))

$$|\mathcal{P}(\nu_n = r) - \mathcal{P}(\nu = r)| \leq \lambda(1 - e^{-\lambda})n^{-1} \qquad (7.5)$$

uniformly on $r \geq 0$ and $0 < \lambda \leq n$. Using the obvious inequality

$$\sum_{r=0}^{\infty} I(r^{[m]} = k) \leq 1$$

and applying (7.4) and (7.5) obtain (7.3).

Estimate (7.3) can be refined as follows

$$\Delta(k, m) = \mathcal{P}(\nu_n^{[m]} = k) - \mathcal{P}(\nu^{[m]} = k) +$$

$$+ \frac{1}{2}E\{I(\nu^{[m]} = k)(\nu^2 - (2\lambda + 1)\nu + \lambda^2)\}n^{-1}.$$

Theorem 7.2. *For all $0 < \lambda \le n$, $1 \le m \le n$ and uniformly in k, the following estimate is true*

$$|\Delta(k,m)| \le 4\lambda^2(1 - e^{-\lambda})n^{-2}. \tag{7.6}$$

Proof. First show that

$$\Delta(k,m) = \sum_{r=0}^{\infty} I(r^{[m]} = k)\Delta(r,1). \tag{7.7}$$

For this purpose consider

$$E\{I(\nu^{[m]} = k)(\nu^2 - (2\lambda + 1)\nu + \lambda^2)\} =$$

$$= \sum_{r=0}^{\infty} I(r^{[m]} = k)(r^2 - (2\lambda + 1)r + \lambda^2)\mathcal{P}(\nu = r).$$

Here

$$(r^2 - (2\lambda + 1)r + \lambda^2)\mathcal{P}(\nu = r) =$$
$$= E\{I(\nu = r)(\nu^2 - (2\lambda + 1)r + \lambda^2)\}.$$

With substitution of these formulas into the definition of the variable $\Delta(k,m)$ obtain (7.7). Then, from (7.7), obtain the inequality

$$|\Delta(k,m)| \le \sum_{r=0}^{\infty} I(r^{[m]} = k)|\Delta(r,1)|. \tag{7.8}$$

According to the theorem from (Barbour et al. (1984))

$$\Delta(k,1) \le 4\lambda^2(1 - e^{-\lambda})n^{-2} \tag{7.9}$$

uniformly for all $r = 0, 1, 2\ldots$ and for all $0 < \lambda \le n$.

Combining (7.5), (7.8) and (7.9), obtain (7.6).

7.2 Rate of Gaussian approximation

Consider a sample ξ_j, $1 \le j \le n$, of independent identically distributed random variables with

$$E\xi_j = \mu,$$
$$E(\xi_j - \mu)^2 = \sigma^2 > 0.$$

Considering the permanent of the sampling matrix

$$A_n^m(\xi) = [a_{ij} = \xi_j;\ 1 \le i \le m,\ 1 \le j \le n]$$

introduce the distributions

$$F_n(x) = \mathcal{P}(\sqrt{n}(n^{-[m]}\text{Per } A_n^m(\xi) - \mu^m)/m\mu^{m-1}\sigma < x)$$

where $\mu \neq 0$ and

$$F(x) = \mathcal{P}(\tau < x),$$

where τ is the standard Gaussian random variable.

As was shown in chapter 4,

$$\sup_x |F_n(x) - F(x)| \to 0 \tag{7.10}$$

as $n \to \infty$, under the condition $mn^{-1/2} \to 0$.

The rate of convergence in (7.10) is given under the same conditions on m and n in the next theorem.

Theorem 7.3. *Let the following conditions*

$$\mu \neq 0,$$

$$\beta = E|\xi_1 - \mu|^3 < \infty \tag{7.11}$$

be satisfied.

Then, for all $2 \leq m \leq \sqrt{n}$, *the following estimate is true*

$$\sup_x |F_n(x) - F(x)| \leq$$

$$\leq \left(\frac{48\sqrt{2}}{\pi\sqrt{\pi}} + \frac{24\sqrt{3}}{\sqrt{\pi}}\right)\frac{\beta}{\sigma^3}\frac{1}{\sqrt{n}} +$$

$$+\left(1 + \left(\frac{\sqrt{2}}{\sqrt{\pi}} + \frac{3}{2}\right)\frac{\sigma}{\mu} + \frac{9}{8}\left(\frac{\sigma}{\mu}\right)^2\right)\frac{m}{\sqrt{n}} +$$

$$+\frac{1}{6}\left(\frac{\sigma}{\mu}\frac{m}{\sqrt{n}}\right)^2 \exp\left(\left(\frac{\sigma}{\mu}\frac{m}{\sqrt{n}}\right)^2\right). \tag{7.12}$$

Proof. According to decomposition formula (1.32)

$$n^{-[m]}\text{Per } A_n^m(\xi) = \sum_{c=0}^{m} m^{[c]}n^{-[c]}\mu^{m-c} \text{ Per } A_n^c(\xi - \mu)/c!. \tag{7.13}$$

Denote

$$S_n = \frac{1}{\sigma\sqrt{n}}\sum_{j=1}^{n}(\xi_j - \mu),$$

$$\delta_n = \frac{m-1}{\sigma\mu(n-1)\sqrt{n}} \sum_{1\leq i<j\leq n} (\xi_i - \mu)(\xi_j - \mu),$$

$$R_n = \frac{\sqrt{n}}{m\mu^{m-1}\sigma} \sum_{c=3}^{m} m^{[c]} n^{-[c]} \mu^{m-c} \text{Per } A_n^c(\xi - \mu)/c!.$$

Following these notations rewrite (7.13)

$$\sqrt{n} n^{-[m]} \text{Per } A_n^m(\xi)/m\mu^{m-1}\sigma$$

$$= S_n + \delta_n + R_n.$$

Hence

$$F_n(x) = \mathcal{P}(S_n + \delta_n + R_n < x), x \in R.$$

Lemma 7.1. *The following inequality is true*

$$\sup_x |F_n(x) - F(x)| \leq$$

$$\leq \sup_x |\mathcal{P}(S_n + \delta_n < x) - F(x)| +$$

$$+ \frac{1}{\sqrt{2\pi}} \frac{\sigma}{\mu} \frac{m}{\sqrt{n}} + \frac{1}{6} \left(\frac{\sigma}{\mu} \frac{m}{\sqrt{n}} \right)^2 \exp\left(\left(\frac{\sigma}{\mu} \frac{m}{\sqrt{n}} \right)^2 \right). \tag{7.14}$$

Proof. First check that an auxiliary inequality is true

$$\sup_x |F_n(x) - F(x)| \leq$$

$$\leq \sup_x |P(S_n + \delta_n < x) - F(x)| +$$

$$+ \Delta_1 + \Delta_2, \tag{7.15}$$

where

$$\Delta_1 := \sup_x |F(x) - F(x + m\sigma/\mu\sqrt{n})|,$$

$$\Delta_2 := \mathcal{P}(|R_n| \geq m\sigma/\mu\sqrt{n}).$$

Indeed,

$$\begin{aligned} F_n(x) &= \mathcal{P}(S_n + \delta_n + R_n < x) \\ &= \mathcal{P}\{(S_n + \delta_n < x - R_n) \cap (|R_n| \leq m\sigma/\mu\sqrt{n})\} \\ &\quad + \mathcal{P}\{(S_n + \delta_n < x - R_n)\} \cap (|R_n| > m\sigma/\mu\sqrt{n})\}. \end{aligned}$$

First of all, it implies

$$\begin{aligned} F_n(x) &\geq \mathcal{P}\{(S_n + \delta_n < x - R_n) \cap (|R_n| \leq m\sigma/\mu\sqrt{n})\} \\ &\geq \mathcal{P}\{(S_n + \delta_n < x - m\sigma/\mu\sqrt{n}) \cap (|R_n| \leq m\sigma/\mu\sqrt{n})\} \\ &\geq \mathcal{P}(S_n + \delta_n < x - m\sigma/\mu\sqrt{n}) - \mathcal{P}(|R_n| > m\sigma/\mu\sqrt{n}). \end{aligned}$$

and, secondly,

$$\begin{aligned} F_n(x) &\leq \mathcal{P}(S_n + \delta_n < x + m\sigma/\mu\sqrt{n}) \\ &\quad + \mathcal{P}(|R_n| > m\sigma/\mu\sqrt{n}). \end{aligned}$$

(7.15) follows from these results.

Estimate of Δ_1.
According to the property of function $F(x)$,

$$\Delta_1 \leq \frac{1}{\sqrt{2\pi}} \frac{\sigma}{\mu} \frac{m}{\sqrt{n}}. \tag{7.16}$$

Estimate of Δ_2.
According to the Chebyshev inequality,

$$\Delta_2 \leq \frac{\mu^2}{\sigma^2} \frac{n}{m^2} E R_n^2.$$

Notice that

$$\begin{aligned} &E\text{Per } A_n^{c_1}(\xi - \mu)\text{Per } A_n^{c_2}(\xi - \mu) \\ &= \begin{cases} 0, & c_1 \neq c_2, \\ c! n^{[c]} \sigma^{2c}, & c_1 = c_2 = c. \end{cases} \end{aligned}$$

Hence, by definition of R_n,

$$E R_n^2 = \frac{n\mu^2}{m^2\sigma^2} \sum_{c=3}^{m} m^{2[c]}\, n^{-[c]} \sigma^{2c} \mu^{-2c}/c!.$$

Here

$$m^{2[c]}/n^{[c]} \leq m^{2c}/n^c.$$

Then

$$E R_n^2 \leq \frac{n\mu^2}{m^2\sigma^2} \sum_{c=3}^{m} \left(\frac{\sigma}{\mu} \frac{m}{\sqrt{n}}\right)^{2c} /c!$$

$$= \left(\frac{\sigma}{\mu}\frac{\mu}{\sqrt{n}}\right)^4 \sum_{c=0}^{m-3} \left(\frac{\sigma}{\mu}\frac{m}{\sqrt{n}}\right)^{2c} \frac{1}{(c+3)!}$$

$$\leq \frac{1}{6}\left(\frac{\sigma}{\mu}\frac{m}{\sqrt{n}}\right)^4 \sum_{c=0}^{\infty} \left(\frac{\sigma}{\mu}\frac{m}{\sqrt{n}}\right)^{2c} /c!$$

$$= \frac{1}{6}\left(\frac{\sigma}{\mu}\frac{m}{\sqrt{n}}\right)^4 \exp\left(\left(\frac{\sigma}{\mu}\frac{m}{\sqrt{n}}\right)^2\right).$$

As a result obtain

$$\Delta_2 \leq \frac{1}{6}\left(\frac{\sigma}{\mu}\frac{m}{\sqrt{n}}\right)^2 \exp\left(\left(\frac{\sigma}{\mu}\frac{m}{\sqrt{n}}\right)^2\right). \tag{7.17}$$

(7.14) follows from (7.15)-(7.17).

Lemma 7.2. *For all* $2 \leq m \leq \sqrt{n}$, *the following inequalities hold*

$$\sup_x |\mathcal{P}(S_n + \delta_n < x) - F(x)| \leq$$

$$\leq \left(\frac{48\sqrt{2}}{\pi\sqrt{\pi}} + \frac{24\sqrt{3}}{\sqrt{\pi}}\right) \frac{\beta}{\sigma^3} \frac{1}{\sqrt{n}} +$$

$$+ (1 + \frac{1}{\sqrt{2\pi}} \frac{\sigma}{\mu} + \frac{3}{2}\frac{\sigma}{\mu} + \frac{9}{8}\left(\frac{\sigma}{\mu}\right)^2) \frac{m}{\sqrt{n}}. \tag{7.18}$$

Proof. With the same reasoning as that for the deduction of (7.15) find the inequality

$$\sup_x |\mathcal{P}(S_n + \delta_n < x) - F(x)| \leq$$

$$\leq \sup_x |\mathcal{P}(S_n + \delta_{nk} < x) - F(x)| + \Delta_1 + \Delta_3, \tag{7.19}$$

where Δ_1 is defined in (7.15) and

$$\Delta_3 := \mathcal{P}(|\delta_n - \delta_{nk}| > m\sigma/\mu\sqrt{n}),$$

$$\delta_{nk} := \frac{(m-1)}{\sigma\mu(n-1)\sqrt{n}} \sum_{1 \leq i < j \leq k} (\xi_i - \mu)(\xi_j - \mu),$$

where $k = n - m[\sqrt{n}]$ and $[\sqrt{n}]$ denotes the integral part of $\sqrt{n}$.

Estimate of Δ_3.
According to the Chebyshev inequality,

$$\Delta_3 \leq \frac{n}{m^2}\frac{\mu^2}{\sigma^2} E(\delta_n - \delta_{nk})^2.$$

By definition

$$\delta_n - \delta_{nk} = \frac{(m-1)}{\sigma\mu(n-1)\sqrt{n}} \sum_{j=k+1}^{n} (\xi_j - \mu) \sum_{i=1}^{j-1} (\xi_i - \mu).$$

Applying this formula write

$$\begin{aligned} E(\delta_n - \delta_{nk})^2 &= \frac{(m-1)^2}{\sigma^2\mu^2(n-1)^2 n} \sum_{j=k+1}^{n} E(\xi_j - \mu)^2 \\ &\quad \times E\,\Big(\sum_{i=1}^{j-1} (\xi_i - \mu)\Big)^2 \\ &= \frac{(m-1)^2}{\sigma^2\mu^2(n-1)^2 n} \sigma^4 \sum_{j=k+1}^{n} (j-1) \\ &= \frac{\sigma^2}{\mu^2} \frac{(m-1)^2(n-k)(n+k-1)}{2(n-1)^2 n}. \end{aligned}$$

Consequently, when $k = n - m[\sqrt{n}]$ and $\sqrt{n} \geq m \geq 2$,

$$E(\delta_n - \delta_{nk})^2 \leq \frac{\sigma^2}{\mu^2} \Big(\frac{m}{\sqrt{n}}\Big)^3.$$

Hence obtain the estimate

$$\Delta_3 \leq \frac{m}{\sqrt{n}}. \tag{7.20}$$

Then consider the first term in the right-hand side of (7.19). First estimate it by the Esseen inequality

$$\begin{aligned} &\sup_x |\mathcal{P}(S_n + \delta_{nk} < x) - F(x)| \leq \\ &\leq \frac{1}{\pi} \int_{-T}^{T} |E \exp(it(S_n + \delta_{nk})) - e^{-\frac{t^2}{2}}| |t|^{-1} dt + \\ &+ \frac{24}{\pi\sqrt{2\pi}} T^{-1}, \end{aligned} \tag{7.21}$$

where $T = \sqrt{n}\sigma^3/4\beta$. It is obvious that

$$\begin{aligned} |Ee^{it(S_n+\delta_{nk})} - e^{-\frac{t^2}{2}}| &\leq |Ee^{it(S_n+\delta_{nk})} - Ee^{itS_n}| \\ &\quad + |Ee^{itS_n} - e^{-\frac{t^2}{2}}|. \end{aligned} \tag{7.22}$$

Let $\varphi(t) = E \exp(it(\xi_1 - \mu))$, $t \in R$. It can be expressed as

$$Ee^{itS_n} = (\varphi(\frac{t}{\sigma\sqrt{n}}))^n$$

because $\xi_1, \ldots, \xi_n$, are independent identically distributed random variables.

In the domain $|t| \leq T$ by well-known estimates

$$|\varphi(t)| \leq e^{-\sigma^2 t^2/3n}, \tag{7.23}$$

$$|Ee^{itS_n} - e^{-\frac{t^2}{2}}| \leq 16\sigma^{-3}\beta n^{-1/2}|t|^3 e^{-\frac{t^2}{3}}.$$

That is why

$$|Ee^{itS_n} - e^{-t^2/2}| \leq 16\sigma^{-3}\beta n^{-1/2}|t|^3 e^{-\frac{t^2}{3}}.$$

Therefore

$$\begin{aligned}\int_{-T}^{T} |E^{itS_n} - e^{-\frac{t^2}{2}}||t|^{-1}dt &\leq 16\sigma^{-3}\beta n^{-1/2} \int_{-\infty}^{\infty} t^2 e^{-t^2/3}dt \\ &= 24\sqrt{3\pi}\sigma^{-3}\beta n^{-1/2}. \end{aligned} \tag{7.24}$$

Further, applying the formula

$$e^{ix} = 1 + ix - x^2 \int_0^1 (1-u)e^{ixu}du, x \in R,$$

write the representation

$$\begin{aligned} Ee^{it(S_n+\delta_{nk})} - Ee^{itS_n} &= itE(\delta_{nk}e^{itS_n}) \\ &\quad -t^2 \int_0^1 E(\delta_{nk}^2 e^{it(S_n+\delta_{nk}u)})(1-u)du. \end{aligned} \tag{7.25}$$

By definition

$$E(\delta_{nk}e^{itS_n}) =$$

$$= \frac{m-1}{\sigma\mu(n-1)\sqrt{n}} \sum_{1 \leq i_1 < i_2 \leq k} E((\xi_{i_1} - \mu)(\xi_{i_2} - \mu)e^{\frac{it}{\sigma\sqrt{n}}\sum_{j=1}^{n}(\xi_j - \mu)}).$$

From the concept of independence

$$E((\xi_{i_1} - \mu)(\xi_{i_2} - \mu)e^{\frac{it}{\sigma\sqrt{n}}\sum_{j=1}^{n}(\xi_j - \mu)}) =$$

$$= (\varphi(\frac{t}{\sigma\sqrt{n}}))^{n-2}\Big(E\Big((\xi_1 - \mu)e^{\frac{it}{\sigma\sqrt{n}}(\xi_1 - \mu)}\Big)\Big)^2.$$

Consequently,

$$E(\delta_{nk}e^{itS_n}) =$$

$$= \frac{(m-1)k(k-1)}{2\sigma\mu(n-1)\sqrt{n}}(\varphi(\frac{t}{\sigma\sqrt{n}}))^{n-2s}\Big(E\Big((\xi_1 - \mu)e^{\frac{it}{\sigma\sqrt{n}}(\xi_1 - \mu)}\Big)\Big)^2.$$

According to (7.19)

$$\left|\varphi(\frac{t}{\sigma\sqrt{n}})\right|^{n-2} \le e^{-\frac{t^2(n-2)}{3n}}$$

and, in addition,

$$\left|E\left((\xi_1-\mu)e^{\frac{it}{\sigma\sqrt{n}}(\xi_1-\mu)}\right)\right| \le \frac{\sigma}{\sqrt{n}}|t|.$$

Therefore obtain the inequality

$$|E(\delta_{nk}e^{itS_n})| \le$$

$$\le \frac{(m-1)k(k-1)}{2(n-1)n\sqrt{n}}\left(\frac{\sigma}{\mu}\right)t^2e^{-\frac{t^2}{3}\frac{(n-2)}{n}}.$$

In accordance with this inequality,

$$\int_{-T}^{T}|E(\delta_{nk}e^{itS_n})|dt \le$$

$$\le \frac{(m-1)k(k-1)}{2(n-1)n\sqrt{n}}\frac{\sigma}{\mu}\int_{-\infty}^{+\infty}dt t^2e^{-\frac{t^2}{3}\frac{n-2}{n}} \le$$

$$\le \frac{3}{2}\frac{\sigma}{\mu}\frac{m}{\sqrt{n}}. \qquad (7.26)$$

Further, applying the concept of independence, conclude that

$$E(\delta_{nk}^2e^{it(S_n+\delta_{nk}u)}) \le$$

$$= (\varphi(\frac{t}{\sigma\sqrt{n}}))^{n-k}E\left(\delta_{nk}^2e^{it\delta_{nk}u}\exp\left(\frac{it}{\sigma\sqrt{n}}\sum_{j=1}^{k}(\xi_j-\mu)\right)\right).$$

Hence, in the domain $|t| \le T$,

$$|E(\delta_{nk}^2e^{it(S_n+\delta_{nk}u)})| \le E\delta_{nk}^2e^{-\frac{t^2}{3}\frac{(n-k)}{n}}$$

and

$$E\delta_{nk}^2 = \frac{(m-1)^2k(k-1)}{2(n-1)^2n}\frac{\sigma^2}{\mu^2}.$$

Then

$$\int_0^1 du(1-u)\int_{-T}^{T}dt|t||E(\delta_{nk}^2e^{it(S_n+\delta_{nk}u)})| \le$$

$$\le \frac{(m-1)^2k(k-1)}{4(n-1)^2n}\left(\frac{\sigma}{\mu}\right)^2\int_{-\infty}^{\infty}dt|t|e^{-\frac{t^2}{3}\frac{(n-k)}{n}} \le$$

$$= \frac{(m-1)^2k(k-1)}{4(n-1)^2n}\left(\frac{\sigma}{\mu}\right)^2\frac{3n}{n-k} \le$$

$$\leq \frac{9}{8}\left(\frac{\sigma}{\mu}\right)^2 \frac{m}{\sqrt{n}}. \tag{7.27}$$

Combining (7.21), (7.22) and (7.24)-(7.27) obtain

$$\sup_x |\mathcal{P}(S_n + \delta_{nk} < x) - F(x)| \leq$$

$$\leq \left(\frac{48\sqrt{2}}{\pi\sqrt{\pi}} + \frac{24\sqrt{3}}{\pi}\right)\frac{\beta}{\sigma^3}\frac{1}{\sqrt{n}} + \left(\frac{3}{2}\frac{\sigma}{\mu} + \frac{3}{2}\left(\frac{\sigma}{\mu}\right)^2\right)\frac{m}{\sqrt{n}}.$$

Formulas (7.14) and (7.18) imply (7.12).

Chapter 8

Dependent variables

In this chapter arrays of dependent random variables such as martingale differences, bootstrap variables and order statistics are considered like the elements of matrices.

8.1 Martingale permanent

Let $\xi_j^{(n)}$, $j \geq 1$ be a set of martingale differences defined on the basic probability space $(\Omega, \mathcal{F}, \mathcal{F}_j^{(n)}, P)$ and adapted to the filtration $\mathcal{F}_j^{(n)}$, $j \geq 0$:

$$E|\xi_j^{(n)}| < \infty,$$

$$E[\xi_j^{(n)}|\mathcal{F}_{j-1}^{(n)}] = 0, \quad j \geq 1.$$

Sampling permanent.
For the sampling matrix

$$A_n^m(\xi^{(n)}) := [a_{ij} = \xi_j^{(n)}; \;\; 1 \leq i \leq m, \; 1 \leq j \leq n], \tag{8.1}$$

consider the sampling permanent Per $A_n^m(\xi^{(n)})$.

Theorem 8.1. *Let the martingale differences $(\xi_j^{(n)}, \mathcal{F}_j^n, 1 \leq j \leq n), n \geq 1$ satisfy the conditions*

$$\max_{1 \leq j \leq n} |\xi_j^{(n)}| \xrightarrow{L^1} 0, \tag{8.2}$$

$$\sum_{j=1}^n \left(\xi_j^{(n)}\right)^2 \xrightarrow{P} 1. \tag{8.3}$$

Then the weak convergence

$$\text{Per } A_n^m(\xi^{(n)}) \Rightarrow H_m(\tau), \quad \text{as} \quad n \to \infty, \tag{8.4}$$

takes place for each fixed m, where H_m is the Hermite polynomial of the m-th degree and τ is the standard normal random variable.

Proof. Denote

$$\tau_{ns} = \sum_{j=1}^{n} \left(\xi_j^{(n)}\right)^s, \quad s = 1, 2, \ldots .$$

Under conditions (8.2) and (8.3) (see, for example, Borovskikh et al., 1988d)

$$\tau_{n1} \Rightarrow \tau \, . \tag{8.5}$$

Condition (8.3) coincides with the relation

$$\tau_{n2} \xrightarrow{P} 1. \tag{8.6}$$

Consider the case $s \geq 3$. From the definition of τ_{ns} it is seen that

$$\begin{aligned} |\tau_{ns}| &\leq \sum_{j=1}^{n} |\xi_j^{(n)}|^s \\ &\leq (\xi_n^*)^{s-2} \tau_{n2}, \end{aligned}$$

where

$$\xi_n^* = \max_{1 \leq j \leq n} |\xi_j^{(n)}|.$$

So, under conditions (8.2) and (8.3), and for $s \geq 3$,

$$\tau_{ns} \xrightarrow{P} 0. \tag{8.7}$$

Represent Per $A_n^m(\xi)$ by means of formula (4.33) as

$$\begin{aligned} \text{Per } A_n^m(\xi) &= H_m(\tau_{n1}, \tau_{n2}) \\ &\quad + \sum_{s=3}^{m} P_s(\tau_{n3}, \ldots, \tau_{ns}) H_{m-s}(\tau_{n1}, \tau_{n2}) m^{[s]}. \end{aligned} \tag{8.8}$$

By virtue of (8.5) and (8.6) it is noted in (8.8) that

$$H_m(\tau_{n1}, \tau_{n2}) \Rightarrow H_m(\tau),$$

and, for $s \geq 3$,

$$P_s(\tau_{n3}, \ldots, \tau_{ns}) \xrightarrow{P} 0$$

due to (8.7). Consequently, in (8.8),

$$\sum_{s=3}^{m} P_s(\tau_{n3}, \ldots, \tau_{ns}) H_{m-s}(\tau_{n1}, \tau_{n2}) m^{[s]} \xrightarrow{P} 0 \tag{8.9}$$

under conditions (8.2) and (8.3). It shows that (8.4) follows from (8.8).

Theorem 8.2. *Let $(\xi_j^{(n)}, \mathcal{F}_j^{(n)}, j \geq 1)$ be the sequence of the series of square integrable martingale differences for which the next conditions are satisfied*

$$\max_{1 \leq j \leq n} |\xi_j^{(n)}| \xrightarrow{P} 0 , \tag{8.10}$$

$$\sup_n E \left(\max_{1 \leq j \leq n} |\xi_j^{(n)}| \right)^2 < \infty , \tag{8.11}$$

$$\sum_{j=1}^{n} (\xi_j^{(n)})^2 \xrightarrow{P} \eta^2 > 0 , \tag{8.12}$$

where either η is $\mathcal{F}_0 = \bigcap_{n=1}^{\infty} \mathcal{F}_0^{(n)}$-measurable or $\mathcal{F}_j^{(n)} \subset \mathcal{F}_j^{(n+1)}$ for each n and j. Then, if $n \to \infty$,

$$\text{Per } A_n^m(\xi) \Rightarrow \eta^m H_m(\tau) , \tag{8.13}$$

where the random variables τ and η are independent.

Proof. Under conditions (8.10)-(8.12) (Borovskikh et al., 1988d)

$$\tau_{n1} \Rightarrow \tau\eta. \tag{8.14}$$

Condition (8.12) coincides with the relation

$$\tau_{n2} \xrightarrow{P} \eta^2 . \tag{8.15}$$

So, under conditions (8.10)-(8.12), from (8.14) and (8.15) obtain

$$H_m(\tau_{n1}, \tau_{n2}) \Rightarrow H_m(\tau\eta, \eta^2).$$

In addition $H_m(\tau\eta, \eta^2) = \eta^m H_m(\tau)$. Obviously, conditions (8.10) and (8.12) imply (8.7). Therefore, relation (8.9) holds. Consequently, statement (8.13) follows from representation (8.8).

Multiple sampling permanent.

Consider a multiple sampling matrix

$$A_n^m(\xi_1^{(n)}, \ldots, \xi_k^{(n)}) = \begin{bmatrix} A_n^{m_1}(\xi_1^{(n)}) \\ \vdots \\ A_n^{m_k}(\xi_k^{(n)}) \end{bmatrix} ,$$

where $\xi_s^{(n)} = (\xi_{sj}^{(n)}, 1 \leq j \leq n), 1 \leq s \leq k$, besides $\xi_1^{(n)}, \ldots, \xi_k^{(n)}$ are independent and, for the fixed s random sequence, $(\xi_{sj}^{(n)}, \mathcal{F}_{sj}^{(n)}, j \geq 1)$ is the martingale difference.

Theorem 8.3 *Let, for each $s = 1, \ldots, k$, the martingale differences $(\xi_{sj}^{(n)}, \mathcal{F}_{sj}^{(n)})$ satisfy the conditions*

$$E(\max_{1\le j\le n}|\xi_{sj}^{(n)}|)^2 \xrightarrow{P} 0 , \tag{8.16}$$

$$\sum_{j=1}^{n}(\xi_{sj}^{(n)})^2 \xrightarrow{P} 1, \tag{8.17}$$

$$\sum_{j=1}^{n} E(\xi_{sj}^{(n)})^2 \longrightarrow 1. \tag{8.18}$$

Then the weak convergence

$$\text{Per } A_n^m(\xi_1^{(n)},\dots,\xi_k^{(n)}) \Rightarrow \prod_{s=1}^{k} H_{m_s}(\tau_s) , \tag{8.19}$$

where $\tau_1,\dots,\tau_k$ are independent standard normal random variables, is true.

Proof. First, by analogy with (4.40), write the representation

$$\text{Per } A_n^m(\xi_1^{(n)},\dots,\xi_k^{(n)}) = \frac{m_1!\dots m_k!}{(2\pi i)^k}\oint\dots\oint\prod_{s=1}^{k}\frac{dz_s}{z_s^{m_s+1}}$$

$$\times\exp(\sum_{s_1=1}^{k} z_{s_1}\tau_n(s_1) - \frac{1}{2}\sum_{s_1=1}^{k}\sum_{s_2=1}^{k} z_{s_1}z_{s_2}\tau_n(s_1,s_2))$$

$$\times\exp(Q_n(\xi_1^{(n)},\dots,\xi_k^{(n)};z_1,\dots,z_k)), \tag{8.20}$$

where

$$Q_n(\xi_1^{(n)},\dots,\xi_k^{(n)};z_1,\dots,z_k) =$$

$$=\sum_{r=3}^{m}(-1)^{r-1}r^{-1}\sum_{s_1=1}^{k}\dots\sum_{s_r=1}^{k} z_{s_1}\dots z_{s_r}\tau_n(s_1,\dots,s_r),$$

$$\tau_n(s_1,\dots,s_r) = \sum_{j=1}^{n}\xi_{s_1j}^{(n)}\dots\xi_{s_rj}^{(n)}.$$

Consider now $\tau_n(s_1,\dots,s_r)$. Under conditions (8.16) and (8.17) for $r=1$ (Borovskikh et al., 1988d)

$$\tau_n(s_1) \Rightarrow \tau_{s_1}.$$

For $\tau_n(s_1,s_2)$ with $s_1=s_2$ and under condition (8.17)

$$\tau_n(s_1,s_1) \xrightarrow{P} 1 .$$

If $s_1 \ne s_2$ then, for $\tau_n(s_1,s_2)$,

$$E\tau_n^2(s_1, s_2) = \sum_{j=1}^{n} E(\xi_{s_1 j}^{(n)})^2 E(\xi_{s_2 j}^{(n)})^2$$

$$\leq E(\max_{1\leq j\leq n} |\xi_{s_1 j}|)^2 \sum_{j=1}^{n} E(\xi_{s_2 j}^{(n)})^2$$

So, under conditions (8.16) and (8.18) for $s_1 \neq s_2$,

$$\tau_n(s_1, s_2) \xrightarrow{P} 0.$$

In the case $r \geq 3$,

$$|\tau_n(s_1, \ldots, s_r)| \leq \max_{1\leq j\leq n} |\xi_{s_3 j}^{(n)} \ldots \xi_{s_r j}^{(n)}| \sum_{j=1}^{n} |\xi_{s_1 j}^{(n)} \xi_{s_2 j}^{(n)}|.$$

From this, regarding (8.16) and (8.17), it is seen that

$$\tau_n(s_1, \ldots, s_r) \xrightarrow{P} 0.$$

Therefore in (8.20), for all $|z_1| \leq \varepsilon, \ldots, |z_k| \leq \varepsilon$,

$$Q_n(\xi_1^{(n)}, \ldots, \xi_k^{(n)}; z_1, \ldots, z_k) \xrightarrow{P} 0.$$

It means that in (8.20), for $n \to \infty$, the weak convergence takes place.

$$\text{Per } A_n^m(\xi_1^{(n)}, \ldots, \xi_k^{(n)}) \Rightarrow$$

$$\Rightarrow \frac{m_1! \ldots m_k!}{(2\pi i)^k} \oint \cdots \oint \prod_{s=1}^{k} \frac{dz_s}{z_s^{m_s+1}} \exp(\sum_{s=1}^{k} z_s \tau_s - \frac{1}{2} \sum_{s=1}^{k} z_s^2). \qquad (8.21)$$

In (8.21) the values of the contour integrals coincide with the product of Hermite polynomials in the right-hand side of (8.19).

Theorem 8.4 *Let* $(\xi_{sj}^{(n)}, \mathcal{F}_{sj}^{(n)})$, *for every* $s = 1, \ldots, k$, *be a square integrable martingale difference for which the next conditions are satisfied:*

$$\max_{1\leq j\leq n} |\xi_{sj}^{(n)}| \xrightarrow{P} 0,$$

$$\sup_n E\left(\max_{1\leq j\leq n} |\xi_{sj}^{(n)}|\right)^2 < \infty,$$

$$\sum_{j=1}^{n} (\xi_{sj}^{(n)})^2 \xrightarrow{P} \eta_s^2 > 0,$$

$$\sup_n \left(\sum_{j=1}^{n} E(\xi_{sj}^{(n)})^2\right) < \infty,$$

where either η_s *is* $\mathcal{F}_{s0} = \bigcap_{n=1}^{\infty} \mathcal{F}_{s0}^{(n)}$*-measurable or* $\mathcal{F}_{sj}^{(n)} \subset \mathcal{F}_{sj}^{(n+1)}$ *for all* n *and* j. *Then, if* $n \to \infty$,

$$\text{Per } A_n^m(\xi_1^{(n)}, \dots, \xi_k^{(n)}) \Rightarrow \prod_{s=1}^{k} \eta_s^{m_s} H_{m_s}(\tau_s), \tag{8.22}$$

where the random vectors $(\eta_1, \dots, \eta_k)$ *and* $(\tau_1, \dots, \tau_k)$ *are independent.*

Proof. As the basis use formula (8.20) and then treat with the integrand, taking into account the method of the proof for (8.2) and (8.3).

8.2 Bootstrap-permanent

Let the random variables $\xi_1, \dots, \xi_N$ take their values in the measurable space $(X, \mathcal{X})$ and have equal distribution P. Denote as

$$\pi_N(B) = N^{-1} \sum_{j=1}^{N} \delta_{\xi_j}(B)$$

the empirical distribution on $(X, \mathcal{X})$. Fix π_N. Then define the random sample $\xi_1^*, \dots, \xi_n^*$ of n elements from the total union of variables with the distribution π_N. So, if the vector $(\xi_1, \dots, \xi_N)$ is fixed, then the random variables $\xi_1^*, \xi_2^*, \dots$ are independent, identically distributed and

$$P(\xi_1^* = \xi_j \mid \xi_1, \dots, \xi_N) = N^{-1},\ j = 1, \dots, N.$$

It means that sampling with replacement is realized, and the sample

$$\xi_1^*, \dots, \xi_n^*$$

is obtained from the whole totality

$$\{\xi_1, \dots, \xi_N\}.$$

The random sample $\xi_1^*, \dots, \xi_n^*$ is called *the bootstrap-sample* and the random permanent based on $\xi_1^*, \dots, \xi_n^*$ is *the bootstrap-permanent.* Henceforth, the symbols P^* and E^* will denote the probability and mathematical expectation with respect to measure π_N.

Sampling bootstrap permanent.
For the sampling matrix

$$A_n^m(\xi^*) = [a_{ij} = \xi_j^*; 1 \le i \le m,\ 1 \le j \le n]$$

consider the sampling bootstrap permanents Per $A_n^m(\xi^*)$.

Theorem 8.5 *The generating bootstrap sampling function*

$$\Pi_n^*(z) := \sum_{m=0}^{n} \frac{z^m}{m!} \text{Per } A_n^m(\xi^*)$$

has the following representation:

$$\Pi_n^*(z) = \prod_{j=1}^{N} (1 + z\xi_j)^{\nu_j} , \tag{8.23}$$

with the integer valued random vector $(\nu_1, \ldots, \nu_N)$, which has multinomial distribution with parameters $(n; N^{-1}, \ldots, N^{-1})$ when the vector $(\xi_1, \ldots, \xi_N)$ is fixed.

Proof. By definition,

$$\Pi_n^*(z) = \prod_{j=1}^{n} (1 + z\xi_j^*) . \tag{8.24}$$

Fix $\xi_1, \ldots, \xi_N$. The distribution of the bootstrap-sample $\xi_1^*, \ldots, \xi_n^*$ is concentrated on $\{\xi_1, \ldots \xi_N\}$. Moreover the conditional probability of the event "ξ_1 belongs to the bootstrap sample ν_1 times, ..., ξ_N belongs ν_N times" is defined by the formula

$$\frac{n!}{\nu_1! \ldots \nu_N!} \left(\frac{1}{N}\right)^n , \quad \nu_1 + \nu_2 + \ldots + \nu_N = n,$$

which coincides with the multinomial vector with the parameters

$$(n; N^{-1}, \ldots, N^{-1}).$$

Besides, in (8.24),

$$\prod_{j=1}^{n} (1 + z\xi_j^*) = \prod_{j=1}^{N} (1 + z\xi_j)^{\nu_j},$$

which finishes the proof of (8.23).

Introduce the bootstrap permanent measure Per $A_n^m(B_1, \ldots, B_m)$ as the permanent of the matrix

$$A_n^m(B_1, \ldots, B_m) := [a_{ij} = \delta_{\xi_j^*}(B_i); \ 1 \leq i \leq m, \ 1 \leq j \leq n].$$

The bootstrap symmetric statistics are defined as follows

$$U_n^* = \int_X \cdots \int_X \Phi(x_1, \ldots, x_m) \text{Per } A_n^m(dx_1, \ldots, dx_m), \tag{8.25}$$

where $\Phi : X^m \to R$ is the symmetric function. It is obvious that $U_n^* = \text{Per } A_n^m(\xi^*)$ for the function $\Phi(x_1, \ldots, x_m) = x_1 \ldots x_m$.

Denote

$$\mu = E\xi_1 , \quad \sigma^2 = E(\xi_1 - \mu)^2,$$

$$\begin{aligned}\mu_N &= E^*\xi_1^* \\ &= \frac{1}{N}\sum_{j=1}^{N}\xi_j, \\ \sigma_N^2 &= E^*(\xi_1^* - \mu_N)^2 \\ &= \frac{1}{N}\sum_{j=1}^{N}(\xi_j - \mu_N)^2. \end{aligned} \tag{8.26}$$

Under the strong law of large numbers, for $N \to \infty$,

$$\mu_N \to \mu \quad (P - a.s.),$$

$$\sigma_N^2 \to \sigma^2 \quad (P - a.s.). \tag{8.27}$$

The expression

$$X_n \xrightarrow{P^*} X \quad (P - a.s.)$$

means that, for each $\varepsilon > 0$ and $\min(n, N) \to \infty$,

$$P(|X_n - X| > \varepsilon \mid \xi_1, \ldots, \xi_N) \to 0 \quad (P - a.s.),$$

and the notation

$$X_n \xrightarrow{d^*} X \quad (P - a.s.)$$

implies that for almost all $\xi_1, \ldots, \xi_N$ the conditional distribution of the random variable X_n, under the condition that variables $\xi_1, \ldots, \xi_N$ are fixed, converges to the distribution of the random variable X.

Theorem 8.6 *If* $E|\xi_1|^2 < \infty$, *then the relation*

$$n^{-[m]}\text{Per } A_n^m(\xi^*) - \mu_N^m \xrightarrow{P^*} 0 \quad (P - a.s.) \tag{8.28}$$

is true when m *is fixed and* $\min(n, N) \to \infty$.

Proof. It is necessary to establish that, for each $\varepsilon > 0$ and $\min(n, N) \to \infty$,

$$\begin{aligned} &P^*(|n^{-[m]}\text{Per } A_n^m(\xi^*) - \mu_N^m| > \varepsilon) = \\ &= P(|n^{-[m]}\text{Per } A_n^m(\xi^*) - \mu_N^m| > \varepsilon \mid \xi_1, \ldots, \xi_N) \to 0 \quad (P - a.s.). \end{aligned} \tag{8.29}$$

For the verification of (8.29) it is sufficient to show that

$$E^*(n^{-[m]}\text{Per } A_n^m(\xi^*) - \mu_N^m)^2 \to 0 \quad (P - a.s.). \tag{8.30}$$

Using decomposition formula (1.32) obtain

$$n^{-[m]}\text{Per } A_n^m(\xi^*) - \mu_N^m = \tag{8.31}$$
$$= \sum_{c=1}^m \mu_N^{m-c} m^{[c]} n^{-[c]}\text{Per } A_n^c(\xi^* - \mu_N)/c!.$$

Here

$$E^*(\text{Per } A_n^{c_1}(\xi^* - \mu_N)\text{Per } A_n^{c_2}(\xi^* - \mu_N)) = 0 \ , \quad \text{if } c_1 \neq c_2$$

and

$$E^*(\text{Per } A_n^c(\xi^* - \mu_N))^2 = n^{[c]}\sigma_N^{2c} c!.$$

Therefore, from (8.31),

$$E^*(n^{-[m]}\text{Per } A_n^m(\xi^* - \mu_N^m))^2 = \tag{8.32}$$
$$= \sum_{c=1}^m \mu_N^{2(m-c)} m^{2[c]} n^{-[c]} \sigma_N^{2c}/c!.$$

From (8.27) and (8.32) follows (8.30).

Theorem 8.7 *If* $\mu \neq 0$ *and* $0 < \sigma^2 < \infty$ *then, for* $\min(n, N) \to \infty$,

$$\sqrt{n}(n^{-[m]}\text{Per } A_n^m(\xi^*) - \mu_N^m)/m\mu_N^{m-1}\sigma_N \xrightarrow{d^*} \tau \ (P - a.s.) \tag{8.33}$$

where τ *is the standard normal random variable.*

Proof. Write (8.31) as follows

$$\sqrt{n}(n^{-[m]}\text{Per } A_n^m(\xi^*) - \mu_N^m)/m\mu_N^{m-1}\sigma_N$$

$$= \frac{1}{\sigma_N\sqrt{n}}\sum_{j=1}^n(\xi_j^* - \mu_N) + R_{nN} \ , \tag{8.34}$$

where

$$R_{nN} = \frac{\sqrt{n}}{m\sigma_N}\sum_{c=2}^m \mu_N^{-c+1} m^{[c]} n^{-[c]}\text{Per } A_n^c(\xi^* - \mu_N)/c!.$$

By the central limit theorem for the bootstrap-variables (Bickel et al., 1981)

$$\frac{1}{\sigma_N\sqrt{n}}\sum_{j=1}^n(\xi_j^* - \mu_N) \xrightarrow{d^*} \tau \ (P - a.s.).$$

For R_{nN}, by means of (8.32),

$$E^* R_{nN}^2 = nm^{-2}\sigma_N^{-2}\sum_{c=2}^m \mu_N^{-2c+2} m^{2[c]} n^{-[c]} \sigma_N^{2c}/c!.$$

From here it is seen that

$$E^* R_{nN}^2 \to 0 \ (P - a.s.).$$

Together with (8.34), this proves (8.33).

Theorem 8.8 *If $\mu = 0$, $0 < \sigma^2 < \infty$, m is fixed, $N = n$ and $n \to \infty$, then the relation*

$$\text{Per } A_n^m(n^{-1/2}\sigma_N^{-1}(\xi^* - \mu_N)) \xrightarrow{d^*} H_m(\tau) \ (P - a.s.)$$

is true.

8.3 Order permanent

Let $\xi_1^{(n)} \leq \xi_2^{(n)} \leq \ldots \leq \xi_n^{(n)}$ be the order statistics of the sample $\xi_1, \xi_2, \ldots, \xi_n$. Consider the permanent random measure

$$\text{Per } A_n^m(B_1, \ldots, B_m) = \sum_{\alpha n \leq j_1 \neq \ldots \neq j_m \leq \beta n} \prod_{s=1}^{m} \delta_{\xi_{j_s}^{(n)}}(B_s) \ , \tag{8.35}$$

where $B_s \in \mathcal{X}$ and the numbers α and β are such that $0 \leq \alpha < \beta \leq 1$.

The permanent symmetric statistic over the permanent measure (8.35)

$$U_n^{\alpha\beta}(\Phi) = \int_X \cdots \int_X \Phi(x_1, \ldots, x_m) \text{Per } A_n^m(dx_1, \ldots, dx_m) \tag{8.36}$$

can be represented in the following form

$$U_n^{\alpha\beta}(\Phi) = \sum_{\alpha n \leq j_1 \neq \ldots \neq j_m \leq \beta n} \Phi(\xi_{j_1}^{(n)}, \ldots, \xi_{j_m}^{(n)}). \tag{8.37}$$

The asymptotic normality of $U_n^{\alpha\beta}(\Phi)$ in the form (8.37) was studied (Janssen et al., (1987)) with $0 < \alpha < 1/2$ and $1/2 < \beta < 1$. If in (8.37) $m = 1$, then

$$U_n^{\alpha\beta}(\Phi) = \sum_{j=\alpha n}^{\beta n} \Phi(\xi_j^{(n)}) \tag{8.38}$$

is an L-statistic . Show that permanent approach gives the possibility to study the asymptotics of $U_n^{\alpha\beta}(\Phi)$ in (8.37) for every $m \geq 1$ using the asymptotics of linear sums of form (8.38).

Note that the generating function of permanents in (8.35) has the representation

$$\Pi_n(z_1, \ldots, z_k) = \prod_{j=\alpha n}^{\beta n} (1 + \sum_{s=1}^{k} z_s \delta_{\xi_j^{(n)}}(B_s)) \ . \tag{8.39}$$

Formula (8.39) follows immediately from (1.15). The treatment of the expression in the right-hand side of (8.39), by means of methods of previous chapters, shows that the asymptotics affect linear sums of the form (8.38).

Introduce into consideration the functional sampling matrix

$$A_n^m(\xi^{(n)}) = [a_{ij} = C_n(j, \xi_j^{(n)}); 1 \leq i \leq m, \quad 1 \leq j \leq n], \tag{8.40}$$

where $C_n : R^2 \to R$ is a measurable function of two variables for each n.

For this matrix define the random permanent

$$\text{Per } A_n^m(\xi^{(n)}) = \sum_{1 \leq j_1 \neq \ldots \neq j_m \leq n} \prod_{s=1}^{m} C_n(j_s, \xi_{j_s}^{(n)}). \tag{8.41}$$

If in (8.41) $m = 1$ then obtain the following linear function from order statistics

$$\text{Per } A_n^1(\xi^{(n)}) = \sum_{j=1}^{n} C_n(j, \xi_j^{(n)}) . \tag{8.42}$$

Various problems of asymptotics of random variables of type (8.42) were investigated in a number of sources under different conditions on $C_n(x, y)$ and $\xi_1, \ldots, \xi_n$.

Consider the case when

$$C_n(j, \xi_j^{(n)}) = I(n\alpha \leq j \leq n\beta)\xi_j^{(n)}/n. \tag{8.43}$$

Let

$$F(a) = \alpha, \quad F(b) = \beta,$$

$$\mu = \int_a^b x dF(x),$$

$$\sigma^2 = EX_j^2,$$

where

$$X_j = g(\xi_j) - Eg(\xi_j), \quad j = 1, \ldots, n,$$

$$g(x) = \begin{cases} a, & x < a, \\ x, & a \leq x \leq b, \\ b, & x > b. \end{cases}$$

Theorem 8.9. *Let m be fixed and $n \to \infty$. If $E\xi_1^2 < \infty$ and condition* (8.43) *for random permanent* (8.41) *is satisfied then the weak convergence*

$$\text{Per } A_n^m((\xi^{(n)} - \mu)/\sigma\sqrt{n}) \Rightarrow H_m(\tau) \tag{8.44}$$

occurs, where τ is a standard normal random variable.

Proof. Apply remark 4.5. For this purpose

$$\tau_{nk} = \sigma^{-k} n^{-k/2} \sum_{j=\alpha n}^{\beta n} (\xi_j^{(n)} - \mu)^k, k = 1, 2, \ldots$$

are substituted into (4.33).

For L-statistics

$$\tau_{n1} \Rightarrow \tau, \tag{8.45}$$

according to the central limit theorem , when $n \to \infty$, and

$$\tau_{n2} \to 1, \quad (a.s.) \tag{8.46}$$

according to the strong law of large numbers.

For $k \geq 3$, by the well-known inequality

$$E|\tau_{nk}| \leq C_k (E\xi_1^2)^{k/2} n^{-(k-2)/2}, \tag{8.47}$$

where the positive constant C_k is independent of n and ξ_1.

Consequently, when $3 \leq k \leq m$,

$$\tau_{nk} \xrightarrow{P} 0. \tag{8.48}$$

Applying relations (8.45)-(8.48) in formula (4.33) obtain (8.44).

Theorem 8.10. *If* $\mu \neq 0$ *and* $n \to \infty$,

$$mn^{-1/2} \to \gamma, \quad \gamma > 0,$$

then, under condition (8.43) *for random permanent* (8.41), *the following estimate is valid*

$$n^{-[m]} \text{Per } A_n^m (\xi^{(n)}/\mu^m) \Rightarrow \exp\left(-\frac{1}{2} \left(\frac{\gamma\sigma}{\mu} \right)^2 + \frac{\gamma\sigma}{\mu} \tau \right)$$

when $E\xi_1^2 < \infty$.

Proof. It is sufficient to apply the method of the proof of theorem 4.7, taking into account (8.45)-(8.47).

Bibliographical comments

These notes contain historical knowledge and references to resources where some information was obtained.

Permanent symmetric functions were introduced in 1812 by the mathematician Binet and, in the same year, by Cauchy. The term "permanent" was proposed by Muir (1882).

Avid interest to permanents among mathematicians was excited by the famous Van der Waerden hypothesis (1926) (see also Marcus et al. (1965)): *if A_n^n is a twice stochastic matrix of $n \times n$ dimension then*

$$\text{Per } A_n^n \geq n!/n^n$$

and "=" takes place if and only if A_n^n is the matrix whose every element is equal to $1/n$.

Permanent theory in its theoretical development was expounded in Minc (1978, 1983, 1987). The problem of the calculation of permanents is considered in Bebiano (1982) and Percus (1971). In the papers of Bapat (1988a, 1988b, 1990), Bapat et al. (1989), Gyires (1973) and Vaughan et al. (1972) some problems of probability theory and mathematical statistics are examined in which a key role belongs to permanents, particularly, the questions of multiparameter variants of multinomial distributions and distributions of order statistics of not identically distributed random variables. The survey of Strok (1990) contains interesting papers on permanent polynomials and their applications in graph theory.

Chapter 1.

1.1. The definition and properties of permanents are taken from Minc (1978).

1.2. Simplifying matrices are introduced by the authors because of their further importance. The Laplace formula is given according to Minc (1978).

1.3. Theorem 1.1, about the decomposition of a permanent of a multiple sampling matrix, was proved in Borovskikh and Korolyuk (1991b, 1992).

1.4. Theorem 1.2, about the generating function, was proved in Borovskikh and Korolyuk (1991b) where two methods of the proof of formula (1.15) are given.

1.5. Shift decomposition of permanents formulas are given according to papers of Borovskikh and Korolyuk (1992, 1993a) and Kaneva (1992, 1993). Theorem 1.7, in the case of $m = n$, exists in Minc (1978). Lemma 1.2 is proved in Rubin et al. (1980) by the method of mathematical induction with respect to parameter m.

Chapter 2.

2.1. Weak convergence in theorem 2.1 was justified by the authors.

2.2. Theorem 2.2 is analogous to theorem 2.1 in the case of permanent counting measures.

2.3. Theorem 2.3 exists also in Borovskikh and Korolyuk (1993a).

2.4. Functional Charlier–Poisson polynomials were studied by Ito (1988) in connection with white noise.

2.5. Theorems of this section belong to the authors.

2.6. Equality in the distributions of theorem 2.8 is a generalization of the well-known fact in the case of $m = 1$.

2.7. Statement (2.86) exists in Borovskikh and Korolyuk (1993) as well. The weak convergence of elementary symmetric polynomials to stable laws was considered in Avram et al. (1986) and Seidl (1986).

Chapter 3.

3.1. Theorems 3.1 and 3.2 were established by the authors (see, for example, Borovskikh and Korolyuk (1992, 1993a)).

3.2. Theorem 3.3 can be found in Borovskikh and Korolyuk (1993a).

3.3. The proof of theorem 3.4 is exposed according to Borovskikh and Korolyuk (1993a). Theorem 3.5 is given according to Kaneva (1992).

3.4. Theorem 3.6 belongs to the authors.

Chapter 4.

4.1. Stochastic Kravchuk polynomials were first considered by Korolyuk et al. (1992), Korolyuk (1993) and then Accardi et al. (1993).

4.2. Theorem 4.1 was obtained by the authors.

4.3. Theorems of this section about weak convergence were considered by the authors. The papers of Avram et al. (1987) and Giraitis et al. (1986) were devoted to Appell's polynomials determined by decomposition (4.30).

4.4. Theorem 4.4 is a well-known fact (see Mori (1981), Mori et al. (1982) and Teicher (1988)).

4.5. Theorem 4.5, for a multiple sampling permanent, is proved by the authors (see Borovskikh and Korolyuk (1991b)).

4.6. The particular case of theorem 4.7, under condition $\mathcal{P}(\varepsilon_1 > 0) = 1$, is proved in Es et al. (1986). For other kinds of dependence of m on n, the properties of elementary symmetric polynomials were studied in Es (1986) and Szekely (1982). While proving theorem 4.8, when $k = 2$, the authors follow Borovskikh and Korolyuk (1992) and, for every $k \geq 2$, Kaneva (1992).

4.7. Theorem 4.9, in another formulation, exists in Mori (1981). Theorem 4.10 is a generalization to the multiple sampling case of theorem 4.9. Theorems 4.11 and 4.12 are given according to Borovskikh and Korolyuk (1993a).

Chapter 5.

5.1. The definition of the permanent random measure and all the results of this section belong to the authors.

Originally a permanent random measure appeared in Borovskikh and Korolyuk (1988b). Permanent symmetric statistics, based on the Dirac δ-measure, with values in Hilbert and Banach spaces were studied in Borovskikh (1983, 1984, 1986, 1988, 1991).

5.2. By other methods, Dynkin et al. (1983) and Mandelbaum et al. (1984) studied a normal approximation scheme and Kubo et al. (1989) and Mustafid (1990a, 1990b) studied a Poisson approximation scheme.

In connection with decomposition (5.21) see also Ogura (1972) and Segal et al. (1976).

5.3. Originally the permanent variant of the decomposition method was formulated by Borovskikh and Korolyuk (1991b).

In presentation of this method the authors follow Borovskikh and Korolyuk (1993a).

The Poisson approximation scheme by Stein's method was studied by Barbour et al. (1983, 1984, 1985, 1986, 1987), Berman et al. (1983), Brown et al. (1979) and Silverman (1978).

5.4. This section is built upon the authors' papers (see Borovskikh and Korolyuk (1991a, 1993a)). The asymptotic properties of symmetric statistics and U-statistics, under the conditions $m = m(n) \to \infty$ and $n \to \infty$, were studied by Es (1986), Es et al. (1986), Frees (1989), Halasz et al. (1976) and Szekely (1982).

5.5. Weak convergence (5.93) was proved by Dehling et al. (1990) by other methods.

Chapter 6.

6.1. This section is written according to Kaneva and Korolyuk (1995).

6.2. This section is expounded according to Kaneva (1995) and Kaneva and Korolyuk (1995).

6.3. The presentation is given according to Kaneva (1995).

Chapter 7.

7.1. Theorems 7.1 and 7.2 are consequences of estimations obtained by Barbour et al. (1992).

7.2. Under restrictions on m, like $mn^{-1/2} = O((\ln n)^{-1})$, the estimations (7.12) were obtained by Es et al. (1986) and Friedrich (1989).

Chapter 8.

8.1. Theorems of this section and their proofs exist in Borovskikh and Korolyuk (1993a). The case $k = 1$ was considered in Teicher (1988).

8.2. The generating function of theorem 8.5 was obtained in Borovskikh and Korolyuk (1991b, 1993a). Varied properties of U-statistics of bootstrap variables were studied in Arcones et al. (1992), Bickel et al. (1981), Borovskikh and Korolyuk (1990b, 1990f), Efron (1979), Helmers et al. (1990), Huskova et al. (1993), Knight (1989) and Praestgaard et al. (1993).

8.3. Originally the permanent random measure of order statistics was studied in Borovskikh and Korolyuk (1991b). In connection with analysis of generating function (8.39) the following papers will be useful: Bapat et al. (1989), Griffin et al. (1989), Janssen (1988), Janssen et al. (1987), Lepage et al. (1981), Mijnheer (1988).

Theorems 8.9 and 8.10 are presented according to Borovskikh and Korolyuk (1991b).

Bibliography

Accardi L., Koroliuk D., Korolyuk V.S. (1993). Stochastic Symmetric Polynomials. Centro Matematica Vito Volterra. Universita di Roma II, Preprint No 154, September '93, 1–13.

Arcones M.A., Gine E. (1992). On the bootstrap of U- and V- statistics. *Ann. Statist.* **20**, 655 - 674.

Avram F., Taqqu M.S. (1986). Symmetric polynomials of random variables attracted to an infinitely divisible law. *Probab. Th. Rel. Fields.* **71**, 4, 491 - 500.

Avram F., Taqqu M.S. (1987). Non-central limit theorems and Appell polynomials. *Ann. Probab.* **15**, 2, 767 - 775.

Bapat R.B. (1988a). Discrete multivariate distributions and generalized log-concavity. *Sankhya, Ser.A.* **50**, 1, 98 - 110.

Bapat R.B. (1988b). Multinomial probabilities, permanents and a conjecture of Karlin and Rinott. *Proceedings of the American Mathematical Society.* **102**, 3, 467 - 472.

Bapat R.B. (1990). Permanents in probability and statistics. *Linear Algebra and Appl.* **127**, 3 - 25.

Bapat R.B., Beg M.I. (1989). Order statistics for non-identically distributed variables and permanents. *Sankhya, Ser.A.* **51**, 1, 79 - 93.

Barbour A.D., Eagleson G.K. (1983). Poisson approximation for some statistics based on exchangeable trials. *Adv. Appl. Prob.* **15**, 3, 585 - 600.

Barbour A.D., Eagleson G.K. (1984). Poisson convergence for dissociated statistics. *J. R. Statist. Soc.* **46**, 3, 397 - 401.

Barbour A.D., Eagleson G.K. (1985). Multiple comparisons and sums of dissociated random variables. *Adv. Appl. Prob.* **17**, 8, 147 - 162.

Barbour A.D., Eagleson G.K. (1986). Random association of symmetric arrays. *Stochastic Anal. Appl.* **4**, 2, 239 - 281.

Barbour A.D., Eagleson G.K. (1987). An improved Poisson limit theorem for sums of dissociated random variables. *J. Appl. Prob.* **24**, 3, 586 - 599.

Barbour A.D., Hall P. (1984). On the rate of Poisson convergence. *Math. Proc. Camb. Phil. Soc.* **95**, 3, 473 - 480.

Barbour A.D., Holst L., Janson S. (1992). *Poisson approximation.* Clarendon Press, Oxford.

Bebiano Natalia. (1982). On the evaluation of permanents. *Pacific J. Math.* **101**, 1, 1 - 9.

Berman M., Eagleson G. (1983). A Poisson limit theorem for incomplete symmetric statistics. *J. Appl. Prob.*, **20**, 1, 47 - 60.

Bickel P.Y., Freedman D.A. (1981). Some asymptotic theory for the bootstrap. *Ann. Statist.* **9**, 6, 1196 - 1217.

Billingsley P. (1968). *Convergence of Probabilistic Measures.* Wiley, New-York.

Borovskikh Yu.V. (1983). Asymptotics of U-statistics and von Mises' functionals. *Sov. Math. Dokl.* **27**, 2, 303 - 308.

Borovskikh Yu.V. (1984). On normal approximation in infinite-dimensional spaces. *Sov. Math. Dokl.* **30**, 2, 584 - 553.

Borovskikh Yu.V. (1986). *Theory of U-statistics in Hilbert space.* Preprint #86.78, Institute of Math. Ukr. Acad. of Sci., Kiev. (In Russian).

Borovskikh Yu.V. (1988). *Theory of U-statistics in Banach spaces.* Preprint #88.3, Institute of Math. Ukr. Acad. of Sci., Kiev. (In Russian).

Borovskikh Yu.V. (1991). UB-statistics. *Theor. Prob. Appl.* **36**, 3, 429 - 446.

Borovskikh Yu.V., Korolyuk V.S. (1984a). *Asymptotic Analysis of Distributions of Statistics.* Naukova Dumka, Kiev. (In Russian).

Borovskikh Yu.V., Korolyuk V.S. (1984b). Approximation of non-degenerate U-statistics. *Theory Probab. Appl.* **30**, 3, 439 - 450.

Borovskikh Yu.V., Korolyuk V.S. (1987a). Limiting representation of U-statistics by the multiple Wiener–Ito integral. *Dokl. Akad. Nauk Ukrain.SSR, Ser. A*, 2, 16 - 18. (In Russian).

Borovskikh Yu.V., Korolyuk V.S. (1987b). Weak convergence of U-statistics to a multiple stochastic integral in Hilbert space. *Dokl. Akad. Nauk Ukrain. SSR, Ser. A*, 4, 9 - 13. (In Russian).

Borovskikh Yu.V., Korolyuk V.S. (1988a). Algebraic decomposition and asymptotics of the permanent of a random matrix. In: *Asymptotic and Applied Problems in the Theory of Random Evolution, Institute of Math., Ukrain. Acad. of Sci., Kiev*, 68 - 74. (In Russian).

Borovskikh Yu.V., Korolyuk V.S. (1988b). Asymptotic theory of U-statistics. *Ukr. Math. J.* **40**, 2, 142 - 154.

Borovskikh Yu.V., Korolyuk V.S. (1988c). Rate of convergence of degenerate von Mises' functionals. *Theory Probab. Appl.* **33**, 1, 125 - 135.

Borovskikh Yu.V., Korolyuk V.S. (1988d). *Martingale Approximation.* Naukova Dumka, Kiev. (In Russian).

Borovskikh Yu.V., Korolyuk V.S. (1990a). Rate of convergence in the Central Limit Theorem for UH-statistics. *Ukr. Math. J.* **42**, 8, 930 - 936.

Borovskikh Yu.V., Korolyuk V.S. (1990b). U-statistics of a bootstrap sample. *Ukr. Math. J.* **42**, 10, 1245 - 1248.

Borovskikh Yu.V., Korolyuk V.S. (1990c). UH-statistics. *Sov. Math. Dokl.* **40**, 2, 432 - 435.

Borovskikh Yu.V., Korolyuk V.S. (1990d). UH-statistics with increasing degrees. *Dokl. Akad. Nauk Ukrain.SSR, Ser. A*, 3, 3 - 7. (In Russian).

Borovskikh Yu.V., Korolyuk V.S. (1990e). Random symmetric polynomials with increasing degree. *Dokl. Akad. Nauk Ukrain.SSR, Ser. A*, 4, 10 - 13. (In Russian).

Borovskikh Yu.V., Korolyuk V.S. (1990f). Elementary symmetric bootstrap polynomials. *Dokl. Akad. Nauk Ukrain.SSR, Ser. A*, 5, 9 - 12. (In Russian).

Borovskikh Yu.V., Korolyuk V.S. (1991a). *U-statistics with increasing degrees.* Preprint #91.3, Institute of Math. Ukr. Acad. of Sci., Kiev. (In Russian).

Borovskikh Yu.V., Korolyuk V.S. (1991b). *Random permanents and symmetric statistics.* Preprint #91.12, Institute of Math. Ukr. Acad. of Sci., Kiev. (In Russian).

Borovskikh Yu.V., Korolyuk V.S. (1991c). U-statistics with increasing degrees. *Sov. Math. Dokl.* **42**, 2, 502 - 506.

Borovskikh Yu.V., Korolyuk V.S. (1991d). U-statistics of increasing degrees with asymptotic Poisson distribution. *Dokl. Akad. Nauk Ukrain.SSR, Ser. A*, 4, 29 - 32. (In Russian).

Borovskikh Yu.V., Korolyuk V.S. (1991e). U-statistics of increasing degrees with asymptotic Poisson distribution. *Report BS-R9120 August. Stichting Mathematical Centrum, Amsterdam*, 1 - 4.

Borovskikh Yu.V., Korolyuk V.S. (1991f). Random permanents and symmetric statistics. *VI USSR-JAPAN Symposium on Probability Theory and Mathematical Statistics. Abstr. of Communic., Kiev*, p. 83.

Borovskikh Yu.V., Korolyuk V.S. (1992). Random permanents and symmetric statistics. *Proceedings of the Sixth USSR-JAPAN Symposium on Probability Theory and Mathematical Statistics*, Kiev. World Scientific, 176 - 187.

Borovskikh Yu.V., Korolyuk V.S. (1993a). *Random permanents.* Institute of Mathematics, Kiev. (In Russian).

Borovskikh Yu.V., Korolyuk V.S. (1993b). Random permanents. *Abstracts of communications, v.1 (A - L). Sixth International Vilnius conference on Probability Theory and Mathematical Statistics, Vilnius. Institute of Mathematics and Informatics*, 200 - 201.

Borovskikh Yu.V., Korolyuk V.S. (1994a). Random permanents and symmetric statistics. *Acta Applicandae Mathematicae*, **31**, 1 - 63.

Borovskikh Yu.V., Korolyuk V.S. (1994b). *Theory of U-statistics.* Kluwer Academic Publishers. Dordrecht.

Brown T.S., Silverman B.W. (1979). Rates of Poisson convergence of U-statistics. *J. Appl. Prob.*, **16**, 2, 428 - 432.

Chen L.H.Y. (1975). Poisson approximation for dependent trials. *Ann. Probab.* **3**, 3, 534 - 545.

Dehling H., Denker M., Woyczynski W.A. (1990). Resampling U-statistics using p-stable laws. *J. Multiv. Anal.* **34**, 1, 1 - 13.

Dynkin E.B., Mandelbaum A. (1983). Symmetric statistics, Poisson point-processes and multiple Wiener integrals. *Ann. Statist.* **11**, 3, 739 - 745.

Efron B. (1979). Bootstrap methods: another look at the jack-knife. *Ann. Statist.* **7**, 1, 1 - 26.

Engel D. (1982). The multiple stochastic integral. *Memoirs of the American Mathematical Society.* **38**, 265.

Es A.I.van (1986). On the weak limits of elementary symmetric polynomials. *Ann. Probab.* **14**, 2, 677 - 695.

Es A.I.van, Helmers R. (1986). Elementary symmetric polynomials of increasing order. *Probab. Th. Rel. Fields.* **80**, 1, 21 - 35.

Feinsilver P.I. (1978). *Special Functions, Probability Semi-groups and Hamilton Flows.* Lecture Notes in Mathematics, **696**.

Frees E.W. (1989). Infinite order U-statistics. *Scand. J. Statist. Theor. and Appl.* **16**, 1, 29 - 45.

Friedrich K.O. (1989). A Berry-Esseen bound for functions of independent random variables. *Ann. Statist.* **17**, 1, 170 - 183.

Giraitis L., Surgailis D. (1986). Multivariate Appell Polynomials and the Central Limit Theorem. In: **Eberlein E., Taggu M.S.** (eds). *Dependence in Probability and Statistics.* Birkhäuser. Boston.

Girko V.L. (1990). *Theory of random determinants.* Kluwer Academic Publishers. Dordrecht.

Griffin P.S., Pruitt W.E. (1989). Asymptotic normality and subsequential limits of trimmed sums. *Ann. Prob.* **17**, 3, 1186 - 1219.

Gyires B. (1973). Discrete distribution and permanents. *Publ. Math. Debrecen.* **20**, 1, 93 - 106.

Halasz G., Szekely G.I. (1976). On the elementary symmetric polynomials of independent random variables. *Acta Math. Acad. Sci. Hungar.* **28**, 3-4, 397 - 400.

Helmers R., Janssen P., Serfling R. (1990). Berry-Esseen and bootstrap results for generalized L-statistics. *Scand. J. Statist. Theory and Appl.* **17**, 1, 65 - 77.

Hoeffding W. (1948). A class of statistics with asymptotically normal distribution. *Ann. Math. Statist.* **19**, 3, 293 - 325.

Huskova M., Janssen P. (1993). Consistency of the generalized bootstrap for degenerate U-statistics. *Ann. Statist.* **21**, 4, 1811 - 1823.

Ito K. (1951). Multiple Wiener integral. *J. Math. Soc., Jap.* **3**, 1, 157 - 169.

Ito Y. (1988). Generalized Poisson functionals. *Probab. Th. Rel. Fields.* **77**, 1, 1 - 28.

Ito Y., Kubo I. (1988). Calculus on Gaussian and Poisson white noises. *Nagoya Math. J.* **111**, 1, 41 - 84.

Janssen A. (1988). Uniform convergence of sums of order statistics to stable laws. *Probab. Th. Rel. Fields.* **78**, 2, 261 - 271.

Janssen A., Serfling R., Veraverbeke N. (1987). Asymptotic normality of U-statistics based on trimmed samples. *J. Statist. Plann. and Inference.* **16**, 1, 63 - 74.

Kallenberg O. (1975). *Random Measures.* Akademie-Verlag. Berlin.

Kaneva H. (1992). Decomposition of the permanent of k-dimensional projection matrices. *Dokl. Akad. Nauk Ukrainy.* **11**, 34 - 37.

Kaneva H. (1993). Symmetric statistics and random permanents. *Abstracts of communications, Sixth International Vilnius Conference on Probability Theory and Mathematical Statistics,* Vilnius. Institute of Mathematics and Informatics. **1 (A - L)**, 166 - 167.

Kaneva H. (1995). Random permanents of mixed multiple sampling matrices. *Ukr. Math. J.* **47.**

Kaneva H., Korolyuk V.S. (1995). Random permanents of mixed sampling matrices. *Ukr. Math. J.* **47.**

Knight Keith (1989). On the bootstrap of the sample mean in the infinite variance case. *Ann. Statist.* **17**, 3, 1168 - 1175.

Korolyuk V.S., Strok V.V. (1992) Kravchuk Polynomials in algebraic code theory. *Abstracts of communications of the International Conference devoted to the memory of M.P. Kravchuk,* Kiev. Institute of Mathematics of the Ukrainian Academy of Sciences. 96.

Korolyuk V.S. (1993). Stochastic Kravchuk Polynomials. *Dokl. Acad. Nauk. Ukraine* **6**, 34 - 37.

Kubo I., Mustafid (1989). Law of small numbers and limit theorems of asymmetric statistics. *Abstracts of communications, v.1. Fifth International Vilnius Conference on Probability Theory and Mathematical Statistics, Vilnius. Institute of Mathematics and Informatics*, 292 - 293.

Kwapien S., Woyczynski W.A. (1992). *Random Series and Stochastic Integrals.* Birkhäuser. Boston.

Lepage R., Woodroofe M., Zinn J. (1981). Convergence to a stable distribution via order statistics. *Ann. Probab.* **9**, 624 - 632.

Loeve M. (1960). *Probability Theory.* Van Nostrand, Princeton. New York.

Major P. (1981). *Multiple Wiener-Ito Integrals.* L.N.M. **849**. Springer-Verlag. Berlin.

Mandelbaum A., Taggu M.S. (1984). Invariance principle for symmetric statistics. *Ann. Statist.* **12**, 2, 483 - 496.

Marcus M., Mine H. (1965). Permanents. *Amer. Math. Monthly.* **72,** 577 - 591.

Mathai A.M., Provost S.B. (1992). *Quadratic Forms in Random Variables. Theory and Applications.* Dekker. New York.

Mijnheer Joop. (1988). On the rate of convergence of the sum of the sample extremes. *Probab. Th. Rel. Fields.* **79**, 3, 317 - 325.

Minc H. (1978). *Permanents.* Addison-Wesley. Englewood Cliffs.

Minc H. (1983). Theory of permanents 1978 - 1981. *Linear and Multilinear Algebra.* **12**, 2, 227 - 263.

Minc H. (1987). Theory of permanents 1982 - 1986. *Linear and Multilinear Algebra.* **21**, 1, 109 - 148.

Mori T. (1981). On a Hoeffding-type problem. *Lect. Notes in Statistics.* **8**, 174 - 181.

Mori T.F., Szekely G.J. (1982). Asymptotic behaviour of symmetric polynomial statistics. *Ann. Probab.* **10**, 1, 124 - 131.

Muir T. (1882). On a class of permanent symmetric functions. *Proc. Roy. Soc. Edinburgh.* **11**, 409 - 418.

Muir T. (1902). The generating function of reciprocal of a determinant. *Trans. Royal. Soc. Edinburgh.* **40**, 615 - 629.

Mustafid. (1990a). A limit theorem of symmetric statistics for non-identically distributed independent random elements. *Hiroshima Math. J.* **20**, 1, 159 - 169.

Mustafid. (1990b). The law of small numbers and the limit theorem for symmetric statistics with mixing conditions. *Hiroshima Math. J.* **20**, 1, 171 - 183.

Ogura H. (1972). Orthogonal functionals of the Poisson process. *IEEE Trans. Information Theory.* vol. **IT - 18**, 4, 473 - 481.

Percus J.K. (1971). *Combinatorial methods.* Springer-Verlag, New-York.

Praestgaard J., Wellner J.A. (1993). Exchangeably weighted bootstraps of the general empirical process. *Ann. Probab.* **21**, 4, 2053 - 2086.

Rosinski I., Woyczynski W.A. (1986). On Ito stochastic integration with respect to top-stable motion: inner clock, integrability of sample paths, double and multiple integrals. *Ann. Probab.* **14**, 1, 271 - 286.

Rubin H., Vitale R.A. (1980). Asymptotic distribution of symmetric statistics. *Ann. Statist.* **8**, 1, 165 - 170.

Sazonov V.V. (1981). *Normal Approximation - Some Recent Advances.* L.N.M. **879**. Springer-Verlag. Berlin.

Segal A., Kailath T. (1976). Orthogonal functionals of independent-increment processes. *IEEE Trans. Information Theory.* vol. **IT - 22**, 3, 287 - 298.

Seidl L. (1986). On the problem of limiting distributions of random symmetric functions. *Teor. Ver. i Prim.* **31**, 4, 671 - 684. (In Russian).

Serfling R.J. (1975). A general Poisson approximation theorem. *Ann. Probab.* **3**, 4, 726 - 731.

Serfling R.J. (1984). Generalized L-, M- and R-statistics. *Ann. Statist.* **12**, 1, 76 - 86.

Silverman B.W., Brown T. (1978). Short distances flat triangles and Poisson limits. *J. App. Probab.* **15**, 4, 816 - 826.

Strok V.V. (1990) Permanent polynomials and their application in graph theory. *Permanents: Theory and Applications.* (Editor: Egorychev G. P.) Krasnoyarsk. 73 - 88.

Surgailis D. (1981). On L^2 and non-L^2 multiple stochastic integrals. *Lecture Notes Control. Inf. Sci.* **36**, 212 - 226.

Surgailis D. (1983). On Poisson multiple stochastic integrals and associated equilibrium Markov processes. *Lecture Notes Control. Inf. Sci.* **49**, 233 - 248.

Szekely G.J. (1982). A limit theorem for elementary symmetric polynomials of independent random variables. *Z. Wahrsch. verv. Gebiete.* bf 59, 3, 355 - 359.

Teicher H. (1988). Distribution and moment convergence of martingales. *Probab. Th. Rel. Fields.* **79**, 2, 303 - 316.

Tjur T. (1980). *Probability Based on Random Measures.* Wiley. New York.

Vaughan R.J., Venables W.N. (1972). Permanent expressions for order statistics densities. *J. R. Statist. Soc.* **34**, 2, 308 - 310.

Van der Waerden B.L. (1926). Aufgabe 45. *Iber. Deutsch. Math. Verein.* **35**, 117.

Subject index

Bei Fragen zur Produktsicherheit wenden Sie sich bitte an:
If you have any questions regarding product safety,
please contact:

Walter de Gruyter GmbH
Genthiner Straße 13
10785 Berlin
productsafety@degruyterbrill.com